colofon

Published by
Ministry of Foreign Affairs
P.O. Box 20061
2500 EB The Hague
The Netherlands

Policy Department
Rural and Urban Development Department

Editors
Paul ter Weel (supervisor)
Harry van der Wulp

Photography
FAO Community IPM Programme, Asia
Peter Ooi
Robert Nugent
Paul ter Weel

Graphic production
Publicity Material Division (DVL/VM)

Contributor
Global IPM Facility

Print
Graphic Promotions, Leusden

ISBN
90-5328-228-9

May 1999

Policy and Best Practice Documents are part of the official policy of the
Minister for Development Cooperation and provide up-to-date
background information on topics that are considered important to
the Netherlands' development assistance programme. Policy and Best
Practice Documents address practical issues in different fields of
development and provide guidelines for implementation.

Other available Policy and Best Practice Documents:

table of contents

Contents

preface

This policy and best practice document is about participatory Integrated Pest Management (IPM). In agriculture, IPM is an approach to crop protection based on ecosystem management. Yield losses are kept within economically acceptable margins by creating ecological conditions that suppress the development of pests. Effective IPM replaces conventional chemical pest control and reduces the need for pesticides to a minimum, in several cases even to zero. It makes crop production safer, more sustainable and more cost-effective. IPM is therefore considered a necessary element of sustainable agriculture and food security policies.

Participatory IPM is an approach to agricultural extension that is farmer-driven and has proved very useful in helping farmers to understand and practise IPM. The difference with conventional extension systems is that it empowers farmers to understand the agro-ecosystem of their fields and take their own crop management decisions based on knowledge they attain first-hand by learning through experimentation. Participatory IPM is regarded as the best option for the sustainable, large-scale introduction of IPM in developing countries. Generally speaking, participatory IPM training is much broader than pest control and, for instance, also involves nutrient and water management. It is the start of a permanent learning and empowerment process.

This document aims to assist staff with various professional background at Netherlands Embassies in developing countries who may be involved in:
- identification, formulation, appraisal or review of IPM and crop protection projects, as well as broader agricultural, environmental or rural development projects that may have a potential for inclusion of participatory IPM components;
- preparation of policy or programme documents, such as annual planning documents for the Netherlands bilateral aid programme; and responding to requests for pesticides.

The benefits of participatory IPM extend further than crop protection. Besides sustainable agriculture, food security and environmental protection, it also contributes to poverty alleviation, rural development and enhancement of the position of rural women. This document may therefore also be of interest to the Netherlands staff working in one of these fields. Further, it aims to assist governments, NGOs and international agencies, which seek Dutch development assistance in the field of sustainable agriculture.

This document is intended as a brief and practical introduction to participatory IPM. It summarises problems caused by pesticide use in agriculture and provides lessons learned from the introduction and application of IPM. It does not attempt to give a comprehensive overview of the present state of crop protection and crop management. Instead, an effort has been made to focus on those issues that are likely to come up in the daily practice of embassy staff. For more detailed technical information, reference is made to the literature listed in Appendix 2.

This document on participatory IPM links up with four other policy documents published by the Netherlands Ministry of Foreign Affairs, namely: 'Sustainable Irrigated Agriculture', 'Sustainable Land Use', 'Water Supply and Sanitation in Developing Countries' and 'Water for the Future: Integrated Water Resources Management'. Together these documents contain an overview of the sustainable use of land and water within the context of the Netherlands' development cooperation.

The text is illustrated with quotes from Indonesian farmers who participated in Farmer Field Schools on IPM. Source: Community IPM, Six cases from Indonesia; FAO Technical Assistance; Indonesian National IPM Program, 1998.

Throughout this document gender-neutral terms, such as 'farmers', 'extension staff' and 'embassy staff', refer to both men and women.

Policy and Best Practice Document 3

LIBRARY & INFORMATION SERVICES
MEDWAY LIBRARY 13/07/06

This book must be returned or renewed by the last date stamped below, and may be recalled earlier if needed by other readers. Fines will be charged as soon as it becomes overdue.

TELEPHONE RENEWALS 0181 331 9617

1 5 JUN 2006

- 9 APR 2008

Participato

Published by the

the
UNIVERSITY
of
GREENWICH

Paul ter Weel and Harry van der Wulp

summary

Section 1

Introduction
Integrated Pest Management (IPM) is an approach to crop protection based on ecosystem management, which maintains the natural equilibrium and as such reduces the risk of damage by pests. Participatory IPM is regarded as the most effective option for the large-scale, sustainable introduction of IPM. It is a process through which farmers improve their understanding of the agro-ecosystem and develop a capacity to take well-informed and independent decisions on how to best manage their crop. Farmers learn to design and conduct field experiments to find solutions to specific plant production problems, such as pests, soil fertility or water use.

In over 20 countries, participatory IPM programmes have invariably demonstrated that pesticide use can be drastically reduced while maintaining production levels and increasing farmers' profits. As a result, participatory IPM is now high on the agenda of many farming communities, national and local governments, donors and aid agencies. There are many challenges and opportunities to initiate, expand or consolidate participatory IPM programmes.

Section 2

Policy and action plan
Participatory IPM is supported by several prominent international policy documents, including UNCED Agenda 21, the Convention on Biological Diversity, the OECD/DAC Guidelines on Pest and Pesticide Management and the Rome Declaration on World Food Security. The Netherlands Development Cooperation regards participatory IPM as a valuable concept that contributes to several development objectives. The Netherlands will continue to support IPM programmes through various forms of assistance, including activities that create awareness and enhance policy reform.

Section 3

Lessons learned
Pesticides have often become the dominant response to pest problems resulting from agricultural intensification. Such reliance on pesticides has proven increasingly unsustainable. Much of current pesticide use is unnecessary, unsafe, counter-productive and economically unjustifiable and destabilises agricultural production. In addition, pesticide use generally has a range of negative effects on human health and the environment. Aid involving pesticide supplies often encourages unsustainable agricultural practices.

The effective introduction of IPM requires a conducive policy environment that does not encourage the use of pesticides. Participatory IPM is applicable to different agricultural production systems and a wide variety of crops. It contributes to community development and helps enhance the position of women.

Section 4

Guidelines for support for participatory IPM
The main ways in which donors can support participatory IPM include:
1. assistance in the establishment and development of national IPM programmes;
2. assistance to participatory IPM projects and studies conducted by NGOs;

3. incorporation of participatory IPM components in ongoing or future projects or programmes on agriculture, rural development, community development, poverty alleviation, etc.

The development of national IPM programmes usually involves the following four stages, which each have their own specific characteristics: awareness-building; development; capacity building; consolidation. Promoting a favourable policy environment and reforming plant protection extension services and research agendas are important elements of national IPM programmes.

introduction

1.1

Brief history of the development of IPM

From the 1950s onward, agricultural production was rapidly intensified through irrigation and the introduction of high-tech packages of selected varieties, fertilisers and pesticides. Production became industrialised and yields increased. Expectations were high: agricultural production seemed to have been brought under control through technological solutions, pests could be controlled by pesticides and food shortages would soon belong to the past.

During the 1960s, it gradually emerged that the situation was different from what had been hoped for. Intensive use of pesticides developed resistance and induced new pest outbreaks which could devastate a crop. Ecosystems were severely disrupted and side-effects associated with intensive pesticide use became apparent. Broad international concern arose about the long-term effects on human health and the environment. Rachel Carson's book "Silent Spring", which described these effects, became an important eye-opener to the public.

The concept of Integrated Pest Management (IPM) was developed in response to these negative implications of intensive chemical pest control. Initially, IPM concentrated on the development and introduction of spraying thresholds. Later, non-chemical control methods were integrated with limited and selective pesticide use. Modern IPM is based on ecosystem management. Yield losses are kept within economically acceptable margins by creating ecological conditions that suppress the development of pests. Researchers working with a wide range of crops around the world proved the technical feasibility of IPM and important successes were achieved in the field. But until the 1980s the challenge of large-scale implementation in developing countries, where small-scale farmers account for the bulk of agricultural production, appeared to be insurmountable. The general belief was that IPM was too complicated for illiterate farmers and that more research was needed before effective IPM techniques could be introduced. Attempts to introduce IPM therefore focused on the development of simple techniques and standardised packages that extension staff could instruct farmers to apply. This system did not work well, mainly because the large variation in field situations was not consistent with standard instructions that had often been developed at test plots of research stations. Generally, farmers did not understand the logic behind the instructions they received and were not able to adjust them to their specific situations. Gradually it became clear that an important bottle-neck for IPM implementation was the centralised formulation of IPM instructions, a top-down extension system and lack of motivation resulting from farmers' lack of involvement.

Projects with NGO involvement had demonstrated that Non-Formal Education methods could assist farmers to understand the ecosystem of their fields and to take crop management decisions based on their own insights. During the 1980s, FAO further developed this concept and assisted a group of Asian countries to establish large-scale national IPM training programmes based on this concept. The new approach retained the responsiveness of the NGO projects but attained the scale of distribution demanded by most governments. It had a strong bottom-up character and became known as Participatory IPM or Community IPM. The corner-stone of this approach is season-long Farmer Field Schools (FFS) where farmers study the agro-ecosystem of their fields and search for and test solutions for specific problems by conducting experiments that they design themselves. Through learning, farmers become experts who are aware of the ecological principles of pest management and capable of making well-informed and independent decisions suited to local conditions. Participatory IPM integrates local farmers' indigenous knowledge with experience gained in other IPM programmes and research-based IPM recommendations.

By 1998, participatory IPM was being practised in over 50,000 communities in a large number of countries, mainly in Asia, but also in Africa and Latin America. Most of these communities grow rice, but IPM is also practised by farmers who grow beans, cabbage, cacao, coconut, coffee, cotton, cowpea, cucumber, egg plant, groundnut, maize, mango, okra, peppers, soybean, sweet potato, tea or tomato, etc. The impact of many participatory IPM projects has been documented in detail. The general picture shows a substantial reduction in pesticide use, equal or higher yields and significant increases in farmer profits (Box 1 provides a selection of examples). Participatory IPM is not restricted to certain crops or production systems, and is having a significant impact on plant protection policies and practices. A growing number of countries are establishing large-scale national IPM programmes based on participatory IPM, which is seen as the way forward to consolidate the gains of intensified production and to make production more sustainable, environmentally sound and cost-effective.

Box 1:

Examples of agro-economic benefits achieved with IPM in developing countries[1]

Rice

Indonesia	*IPM/FFS-educated farmers reduced pesticide use to zero, or near-zero. Reduced pesticide use and better fertiliser use increased yields significantly (often 10-20%). Farm-level profits increased significantly but vary per district. Increases of 15-25% seem typical.*
Philippines	*IPM/FFS-educated farmers reduced insecticide application and saved USD 10-15 per hectare per season.*
Vietnam	*IPM/FFS-educated farmers reduced their use of pesticides (mainly insecticides, but also fungicides and herbicides) by 73%. A study of over 1300 villages showed a 4% yield increase in rice and over 20% increase in profits.*
West Africa	*First series of Farmer Field Schools in Ghana, Cote d'Ivoire and Burkina Faso showed savings for rice farmers of over USD 90 per hectare, with yields maintained or increased. Profits increased by over 25%*

Cotton

Pakistan	*FFS-educated farmers reduced pesticide applications from 6 per season to 2, while yields remained the same or slightly increased and profits rose by 20%.*
India	*Pesticide-spraying frequency reduced from 10 to 6, while yields increased from*

	1720 kg/ha to 2050 kg/ha. Net income of farmers increased by USD 123 per hectare.
Sudan	1979-1991, number of insecticide applications per season reduced from 9 to 5; from 1991 to 1997 the number of applications was further reduced to 3, delivering annual savings of over USD 13m.

Vegetables

Philippines	Cabbage: IPM training resulted in farmers reducing applications from over 20 per season to 3, including one treatment with an insect pathogen.
Malaysia	Cabbage: Insecticide sprays reduced from 7-9 to a maximum of 3. Marketable yields up by 5-60%. Profits increased by a factor of 6.
Vietnam	Farmer Field School trials achieved the following reductions in pesticide applications: cabbage from 6 to 1; tomato from 14 (insecticides, fungicides) to zero. Spraying of soybean was reduced from 5 to 1, while profits increased by 20%.
Kenya	Tomato: Preliminary results of a pilot project indicate that tomato farmers managed to reduce pesticide applications by up to 70% while maintaining yields.
Brazil	Soybean: 93% reduction in pesticide use and better yields.
Colombia	Tomato: Pesticide applications reduced from 20-30 to 2-3, saving USD 650 /hectare.

Other crops

Pakistan	Mango: Pesticide applications reduced from 5 to 1; costs of chemical control to farmers reduced by over 90%. Sugar cane: Use of insecticides completely stopped; over USD 2m savings annually in total area of 120,000 ha.
Vietnam	Tea: Farmer Field School trials achieved spraying reductions from 10-12 applications to 4, while increasing yields by 15-40%.
Costa Rica	Banana: Initial 75% reduction with use of thresholds, falling to zero after several years.

1.2   Definition and changing focus of IPM

In 1968, FAO defined **IPM** as: "a pest management system that, in the context of the associated environment and the population dynamics of the pest species, utilises all suitable techniques and methods in as compatible a manner as possible, and maintains the pest population at levels below those causing economically unacceptable damage or loss". Pests include invertebrates (insects, mites, nematodes), plant pathogens (fungi, bacteria and viruses), weeds and vertebrates (rodents and birds).

Some examples of available techniques are: biological control by predators, parasites or insect pathogens; use of pest-resistant crop varieties; adoption of cultural practices that prevent buildup of pests, such as crop rotation, inter-cropping, timing of planting; trapping of pests with trap crops. Selective and judicious use of pesticides is regarded as a last-resort control option. Use of bio-pesticides and semiochemicals is preferred to use of conventional pesticides. Of particular importance is biological control by natural enemies already available in the local ecosystem. Cultural techniques are used to discourage the build-up of pest populations and to maintain optimum levels of natural enemy populations.

The old FAO definition however does not reflect the crucial role of farmers in IPM imple mentation. A farmer must be able to transform and adapt any chosen technology to suit her/his actual ecological and market situation. It is increasingly being recognised that when training farmers to practise IPM the approach and process are more important than the transfer of technology itself. The new term **Participatory IPM** puts emphasis on the process of enabling farmers to attain agro-ecological knowledge as a basis for sustainable production. Participatory IPM enhances ecological awareness and stresses the responsibility of farmers for diagnosing pest problems and actively seeking solutions best suited to the situation in their fields. It integrates local knowledge and is site specific.

The earlier IPM was centrally designed and largely driven by public research institutions and agro-chemical companies, while the later participatory IPM is locally designed and driven by farmers themselves. The earlier IPM focused on pests, while participatory IPM is broader and addresses amongst others soil fertility, crop rotation, as well as pests. Another important difference between the earlier IPM and participatory IPM is that the latter has a strong social component which encourages farmers to organise themselves and thus enhances follow-up and sustainability.

Participatory IPM is also referred to as **Community IPM** because farmers themselves become the initiators, implementers and primary beneficiaries of IPM activities. It enhances individual and collaborative decision-making, farmer confidence, business skills and the development of effective local organisations. It empowers rural communities to take better control of their own situation and decreases their earlier dependence on external services.

1.3   Key issues and challenges

Although participatory IPM programmes have achieved impressive results, there remain many challenges and opportunities to expand and consolidate these programmes and to further improve policies. There are also many requests for assistance or guidance in the establishment and development of new projects.

Assistance is required to help set up and develop IPM projects and national programmes. At the same time, donors and aid agencies should critically review their assistance to other agricultural projects to ensure that these do not continue to reinforce the old bias towards pesticide use as the main method of crop protection.

The key issues and challenges are to:
- establish and/or further develop national IPM programmes that make participatory IPM training available to larger numbers of farmers. Within these programmes,

special attention should be paid to enabling and encouraging farmer initiatives to continue experimenting after their Farmer Field School (FFS) and to spread IPM through their communities;
- further improve the involvement of women by incorporating explicit gender strategies at all stages of project development;
- improve the policy environment for IPM through reform of policies and practices that directly or indirectly provide irrational support for chemical control;
- reform crop protection research and extension schemes so that they give more support to farmer initiatives in IPM;
- enhance coordination and cooperation among governments, NGOs, donors and international organisations to achieve a high degree of coherence among different IPM projects and programmes;
- integrate participatory IPM into policies on sustainable agriculture and link the FFS approach to other development fields, such as soil conservation, community forestry, irrigation management, etc.).

The challenge of reducing pesticide use is well illustrated by the fact that in 1996, the global pesticide market grew by 5.5% to reach a nominal end-user value of USD 30.5 billion[2]. Although developing countries accounted for less than a third of world pesticide sales, they were the fastest growing market segment. In many countries there is a huge potential to reduce pesticide use through IPM. The potential for reductions is greatest with regard to insecticides, but is certainly not restricted to this group of pesticides.

In 1990 the Netherlands, itself one of the world's most intensive users of pesticides, adopted a policy of reducing pesticide use by at least 50% by the year 2000[3].

Vegetable farmer spraying insecticides on cabbage crop, Cambodia, (photo: Robert Nugent)

policy and action plan

2.1 International policy framework

Policies aimed at food security increasingly focus on environmentally sound and sustainable agricultural production. In this regard, the role of IPM as a technical approach, and participatory IPM as a process approach, are widely acknowledged. Several prominent international policy documents reflect this line and call for an international effort to bring participatory IPM to the large numbers of farmers in developing countries. The main policy documents in this respect are:

UNCED Agenda 21

In 1992, the United Nations Conference on Environment and Development (UNCED) assigned an important role to IPM in the agricultural programmes and policies envisaged as part of its Agenda 21. The text explicitly stated that: "Chemical control of agricultural pests has dominated the scene, but its overuse has adverse effects on farm budgets, human health and the environment, as well as on international trade. New pest problems continue to develop. IPM, which combines biological control, host plant resistance and appropriate farming practices and minimises the use of pesticides, is the best option for the future, as it guarantees yields, reduces costs, is environmentally friendly and contributes to the sustainability of agriculture." Regarding the implementation of IPM, Agenda 21 established the following goals for the world community: 1) to improve and implement programmes to put IPM practices within the reach of farmers through farmers networks, extension services and research institutions; 2) not later than 1999, to establish operational and interactive networks among farmers, researchers and extension staff to promote and develop IPM.

OECD/DAC Guidelines on pest and pesticide management

In 1995, the Development Assistance Committee of the Organisation for Economic Cooperation and Development (OECD/DAC) published guidelines on pest and pesticide management as part of a series which seeks to improve and coordinate OECD-member policies to integrate development and environment. In these guidelines, three priority areas are identified for assistance in the field of pest and pesticide management in developing countries: i) promoting the development and application of IPM; ii) strengthening pesticide management capabilities in developing countries; iii) ensuring good practices

Girl transplanting rice, Cambodia (photo: Robert Nugent)

when providing pesticides under aid programmes. The Guidelines call for a shift from pesticide supply to human resource development to facilitate participatory IPM, which is acknowledged as a vital element of sustainable crop protection.

Convention on Biological Diversity
In November 1996, the Conference of Parties to the Convention on Biological Diversity adopted a decision on conservation and sustainable use of agricultural biological diversity. In the preamble to the decision it is recognised that traditional farming communities and their agricultural practices have made a significant contribution to the conservation and enhancement of biodiversity and that these can make an important contribution to the development of environmentally sound agricultural production systems. It is recognised that inappropriate use of, and excessive dependence on agrochemicals have affected biological diversity. Parties to the Convention decided to encourage the development of technologies and farming practices that not only increase productivity, but also arrest degradation as well as reclaim, rehabilitate, restore and enhance biological diversity. These could include organic farming, integrated pest management, biological control and cultural methods.

Rome Declaration on World Food Security and the World Food Action Plan
The FAO World Food Summit was held in 1996 and produced the Rome Declaration on World Food Security and a World Food Summit Plan of Action. Signatories to the Rome Declaration committed themselves to pursue participatory and sustainable agriculture policies and practices. Among other intentions, governments, in partnership with all civil society actors, and with the support of international institutions, agreed to promote the widespread development and use of IPM practices.

Participatory IPM attracts increasing attention from the international donor community. Several donors and development banks have made IPM the focus of their assistance relating to plant protection, and have prepared specific guidelines in support of IPM. Nevertheless many mistakes are still being made in aid involving agricultural intensification activities and input supplies.

In 1995, a group of international organisations founded the Global IPM Facility, which became operational in 1997. Its purpose is to help expand participatory IPM (see box 6).

2.2   The Netherlands' Development Cooperation Policy and IPM

The overall development objectives of Dutch development cooperation policy are poverty alleviation and sustainable economic growth. Within this context much importance is attached to the development of national policies and institutional capacities, public participation, sustainable agriculture and environmental protection.

In 1990, the principles and outline for Dutch development cooperation were laid down in a policy document entitled *"A World of Difference: A new framework for development cooperation in the*

1990s". Among other issues, it mentioned that IPM holds the promise of a solution to the many environmental problems associated with modern agriculture. The document declared the Netherlands' intention to increase its involvement in IPM. This message was repeated in 1993 in a follow-up document entitled "*A World in Conflict*", which, among other issues, elaborated on the tension between production and environmental protection. Sustainable land use, with IPM as an essential element, is seen as a long-term solution to this tension.

This issue of Policy and Best Practice Documents elaborates on the envisaged role for IPM and on recommendations made in the Sectoral Policy Document on Sustainable Land Use, particularly those with regard to integrated crop management and training and education. Further, it addresses several of the concerns raised in the Sectoral Policy Document on Biological Diversity.

Participatory IPM is explicitly more than an environmentally sound approach to plant protection. As illustrated below, it also contributes to development objectives of other aid sectors important to the Netherlands' aid programme.

Sustainable agriculture and food security

IPM adds long-term sustainability to crop production and is therefore an essential element of food security strategies. IPM restores and maintains the agro-ecological balance in intensified agricultural production systems. Healthier production systems have a reduced risk of devastating pest outbreaks. High production levels can be maintained while reducing costly pesticide inputs.

Macro-economics and structural adjustment

Introduction of IPM in important agricultural sectors may lead to substantial savings in hard-currency spending on pesticide imports. FFS-educated rice farmers in Asia save over USD 10 worth of pesticides per ha per season[4]. Nationwide such savings add up to millions of dollars[5]. Abolition of pesticide subsidies under structural adjustment programmes is more easily accepted by governments if IPM programmes have demonstrated that present levels of pesticide use do not contribute to increased food production. In addition, importers increasingly demand products free of pesticide residues. IPM-based production may eventually provide a comparative advantage on export markets. This trend is already noticeable for cotton and fruits.

Rural development: Poverty alleviation, farmer empowerment and community development

Savings on unnecessary pesticide use achieved by IPM generally have a significant impact on the net profits of small-scale farmers. Farmers who participated in IPM Farmer Field Schools (FFS) are more confident about their capabilities. They often organise themselves to further enhance their IPM knowledge and practices after the FFS and become more vocal towards extension services, resulting in more effective interaction with such services. Farming communities increasingly, and successfully, demand that participatory IPM training is included in local development policies and activities.

National policies and institutional capacities

In several countries, the introduction of participatory IPM has led to reforms in agricultural policies and services. IPM activities were integrated in government extension services and investment shifted from input supply to human resource development and building capacity to conduct large-scale IPM field training programmes. Several governments confidently abolished subsidies for pesticides. In many Asian villages IPM became an issue in local politics leading to local government support for IPM activities.

Empowerment of women

A large proportion of farmers in developing countries are women. They are actively involved in most on-farm and off-farm work. Farmer Field Schools have demonstrated a positive impact on the position of participating women. They learn agricultural skills, become more self-confident and better integrated into development activities at the local level.

Environmental protection and biodiversity

Overuse of pesticide has affected populations of birds, fish, frogs, shrimp, bees, earthworms, etc. Reduced pesticide use and better selection of pesticides diminish environmental pollution. Ecosystems disrupted by excessive pesticide use often recover when production systems switch to IPM. Maintaining biodiversity is a pillar of modern IPM, which aims to prevent sudden pest outbreaks by preserving the natural integrity of ecosystems. FFS-educated farmers become custodians of biodiversity[6].

Farmer Field School making posters, Philippines (photo: Paul ter Weel)

Public and occupational health

Reduced pesticide use and better selection of pesticides diminishes direct exposure of farmers to pesticides. Small-scale farmers rarely have access to appropriate information and protective gear to reduce the hazards of handling and applying pesticides. High and low-level poisoning of farmers working with pesticides is a widely occurring phenomenon throughout the developing world. IPM also reduces the risk of unacceptable levels of pesticide residues in consumer products and drinking water.

2.3   Sustainability of participatory IPM

Participatory IPM is knowledge-based. Farmers attain knowledge and understanding of
and insight into the agro-ecosystem of their crops and the economics of pesticide use.
This knowledge enables them to make their own decisions about crop management.
They continue with IPM because they understand what they are doing and calculate what
they are gaining. They are able to adapt their practices to changing circumstances
because they own the process and not just the conclusion of someone else's process, as
used to be the case with more conventional instruction-based extension schemes. Evalu-
ations indicated that farmers value their own experiences and experiments more than
extension recommendations[7].

In many national IPM programmes farmers create their own structures to continue as
groups after the initial FFS field training. Sometimes such groups even intoduce new
crops. Funding of follow-up group activities is generally mobilised using local resources.
Local government or community organisations often get involved. In Indonesia, several
thousands of FFS have been funded by village - sub-district - or provincial funds. A similar trend is emerging in other countries.
In Asia, investment in national IPM programmes in the 1980s started with donor funding
and was gradually followed by funding from national or local government budgets and
loans from development banks. By mid-1996, the accumulated investment from national
government funds had become twice as high as the total provided over time by donors
and aid agencies. The total amount borrowed from development banks for implementation of participatory IPM was about 20% higher than the total so far provided by donors
and aid agencies[8]. This shift from grant funding to national funding and loans should be
regarded as an important indicator of sustainability.

2.4   Netherlands support for participatory IPM

The Netherlands Development Cooperation regards participatory IPM as a valuable concept that contributes to several development objectives. From 1982, the Netherlands
became one of the main donors to the Intercountry Programme for the Development and
Application of IPM in Rice in South and Southeast Asia. It took part in several major eval-
uations and played an active role in the programme's development . In 1995, a parallel
intercountry programme on vegetable IPM was started with Dutch funding in four Asian
countries.

In 1993, a global IPM meeting enabled key technical and policy staff from African and
Latin American countries to familiarise themselves with the achievements in Asia. This
resulted in a wave of requests from governments in these regions, notably Africa, for
assistance to initiate participatory IPM projects. Positive pilot projects further fuelled this
demand. The Netherlands encouraged the move to Africa and became a contributor to

the Global IPM Facility, which was founded to facilitate the further spread and development of participatory IPM (see Box 6).

The Netherlands will continue to support the establishment, development and consolidation of national IPM programmes through project assistance and activities that create awareness and enhance policy reform. Where relevant, participatory IPM field training will be incorporated in broader agricultural projects supported by the Netherlands, such as irrigation projects. In addition, the Netherlands will continue to advocate participatory IPM as a key to sustainable agriculture and food security, at both country level and international level.

lessons learned

3.1   Lessons learned from agricultural intensification and the Green Revolution

Food production needs to increase each year to feed the growing world population. The potential for expanding the area under agricultural cultivation is often limited. Most governments therefore focus on intensifying production in existing agricultural areas.

Farmer Trainer in Takeo province, Cambodia (photo: Robert Nugent)

During the Green Revolution, policies on agricultural intensification commonly involved centrally designed schemes with a high dependency on external inputs comprising specific seed varieties, fertilisers and pesticides. Such schemes often marginalised local and indigenous knowledge and reduced farmers to mere implementers of instructions. Heavily centralised agricultural and extension systems excluded farmers from the decision-making process regarding packages and practices, even though these decisions directly affected their livelihood. Although policies are gradually changing, these practices are still prevalent.

Some of the lessons learned:

- **Intensive use of pesticides often led to crisis situations in which pests became almost uncontrollable and production nearly collapsed.**

In many agricultural production areas the agro-ecological balance has been severely disrupted by intensification programmes that involved heavy reliance on pesticides. Populations of natural enemies of pests vanished as a result of intensive spraying. This accelerated the resurgence of primary pest populations and induced outbreaks of secondary pests. Combined with increased pesticide resistance among pest populations, this often led to crises in which more and more pesticides were used without preventing a decline in production. In several countries, such crises have had major negative economic consequences. Well-known examples are the Brown Plant Hopper crises in rice, the American Bollworm and Whitefly crises in cotton and the Diamondback moth crises in cabbage production. The more pesticides were sprayed, the bigger the outbreaks that followed. Drastic reduction of pesticide use and a shift to IPM-based crop management appeared to be the only long-term solution to what is popularly referred to as the "pesticide treadmill". Similar examples of spiralling pesticide use curbed by IPM exist for various other crops, including bananas, coffee and tomatoes. The main lesson learned is that IPM is an essential component of agricultural intensification.

•   **Often there is no economic justification for intensive pesticide use**
A general bias towards chemical control as the primary mode of crop protection has evolved as a result of: aggressive pesticide marketing; crop protection research overemphasising chemical control because this direction of research attracts additional funding from agro-chemical companies; agricultural extension staff preoccupied with pesticides because they received training from agro-chemical companies and supplement their income with commissions on sales. The result is a general tendency towards excessive use of pesticides. Case-after-case, IPM programmes have demonstrated that common practices of pesticide use are counter-productive, have no economic justification, and are not desirable from a crop protection point of view.

> "I commonly sprayed rice as many as four times. Now I don't spray at all. I have also eliminated spraying my corn."(Muhammad Amanah, IPM farmer, West Lombok, Indonesia)

> "Farmers will spray cabbage as many as 16 times a season or once every three days. IPM farmers have been experimenting and they are now down to four applications per season." (Ms. Tampi, IPM farmer, Banturejo, Ngantang, East Java, Indonesia)

Over the years, farmers have become so accustomed to the use of pesticides that they consider it an obvious and inevitable element of crop production. They have been encouraged to react to visual insect damage without understanding the implications of such damage for their yields and profits. Perceptions of economic damage caused by insects are often based on exaggerated extrapolations of visual crop damage and ignore the crop's capacity to recover from damage and the ecological cost of spraying in terms of increased risk of further and larger pest outbreaks.

Agro-chemical companies, commodity boards, research institutes and extension services have long been promoting calendar-based spraying schemes. Farmers are advised to spray after fixed intervals to prevent the emergence of pests, regardless of the actual pest situation. Farmers have been made to believe that regular spraying minimises the risk of outbreaks. They regard expenditure on pesticides as an insurance premium against crop losses, which they pay to have peace of mind. However, the reality is that calendar-based spraying schemes destroy the natural ecosystem and destabilise production. Many large pest outbreaks have been induced by regular spraying.

In both the developed and the developing world there is a huge potential for pesticide reduction. Present practices of pesticide use urgently require in-depth and case-by-case review. Pesticide use should not automatically be the first control option, as is still often the case. Although some pesticide use may be justifiable, present levels of pesticide use are not needed to feed the growing world population, as is often suggested in promotional materials. Rather, present levels undermine food security by destabilising production, contaminating natural resources and increasing production costs.

- **IPM offers a perspective for sustainable intensification**

IPM has evolved into a mature approach to pest control, suitable for both small- and large-scale implementation in developing countries. Box 1 provides examples of the agro-economic benefits that were achieved by switching to IPM. In all of these cases, production became more cost-effective because unnecessary spending on pesticides was eliminated. Yields often increased and production became more stable. Occupational health hazards for farmers declined with diminishing and more selective pesticide use. The same applied to environmental pollution and public health hazards. But probably the most important result was that the cycle of dependency on pesticides was broken. IPM-based pesticide reductions restored the natural balance of agro-ecosystems and made crops less vulnerable to pests, thereby making production more sustainable.

- **The success of IPM is influenced by government policies on agriculture, environment and public health**

Governments can strongly influence pesticide use and the scope for IPM through legislation, policies and related incentive structures. A favourable policy environment is of great importance to national IPM programmes.

Good pesticide legislation and its effective enforcement help to restrict the availability of highly hazardous and broad spectrum pesticides that are unsuitable for use by small-scale farmers or are incompatible with IPM. Abolition of pesticide subsidies and 'true-costing' of pesticides help avoid irrational use and artificial comparative advantages for production systems based on intensive pesticide inputs. Direct and hidden subsidies for pesticides encourage overuse and act as a brake on innovation and change. Although intended as a means of reducing production risks, experience has shown that pesticide subsidies often have the opposite effect and encourage unsustainable production methods. Subsidised pesticide supply was an important factor in the development of pest control crises in rice and cotton. Examples of indirect subsidies include: preferential import duties and exchange rates for pesticide imports; pesticide sales tax exemption or reduced VAT rates; costs of storage and transport not being included in the price but covered from public budgets; subsidised agricultural credit schemes tied to pesticide use, etc.

Strong environmental and public health policies generally provide disincentives for pesticide use. Maximum residue levels for food crops are an example. Weak policies, on the other hand, may enable continuation of undesirable practices and pesticide misuse. Research and extension programmes may have different approaches to crop protection. Research that largely concentrates on chemical control will contribute less to sustainable development than research on IPM. Extension schemes that concentrate on input supply rather than human resources development will certainly inhibit the spread of IPM. Directions of research and extension are reflected in the policies and budgets of the relevant departments. The lesson learned is that introduction of large-scale IPM programmes should go hand in hand with policy reform to create an enabling policy environment for IPM.

The cost of pesticide use is often also distorted by ignoring hidden costs, such as health costs and loss of labour due to poisoning and costs associated with contamination of ecosystems and drinking water. Contamination of agro-ecosystems may affect agricultural output because of higher risk of pest outbreaks or because of diminished populations of pollinators. Fish and shrimp resources affected by pesticide contamination are another cost factor.

Box 2:

Examples of macro-economic implications of changes in pesticide policies

In 1988, the government of **Indonesia** *abolished pesticide subsidies after it had been demonstrated that current pesticide use in rice induced pest outbreaks. Abolition of subsidies saved the government USD 120 million per year. Major insect pest outbreak frequency declined from 3 (1977-1986) to zero (1987-1996). Outbreaks had cost Indonesia over USD 1.5 billion. National rice production increased by over 15% from 1986 to 1992. The average number of insecticide treatments by rice farmers declined from 4 (1986) to 2.1 (1989); FFS-educated farmers further reduced treatments to near-zero while increasing yields. Farm-level profits per hectare by IPM-educated farmers increased by over USD 10 per season.*

In the **Philippines**, *abolition of subsidies for rice pesticides saved the government over USD 10 million annually.*

The central government in **India** *replaced subsidies on insecticides by a tax and is now gaining USD 60 million annually in savings and revenues, while annual spending on IPM field training is about USD 10 million.[9]*

• **Aid involving pesticide supplies often encourages unsustainable agricultural practices**

Pesticide donations and loans (direct and on-lending) for the procurement of pesticides generally tend to overemphasise the role of chemical control. Donated pesticides are often distributed free of charge or at prices below commercial prices. Such donor subsidies encourage pesticide overuse and abuse, and distort comparative production costs to the disadvantage of IPM-based production methods.

Aid programmes and development bank loans aimed at small-scale farmers still often offer agricultural input packages that include the supply of pesticides or a credit facility to buy pesticides. This is particularly the case with loans channelled through produce boards or cooperatives, aimed at a specific crop (e.g. cotton, tobacco, coffee, rice). Often these are standard packages that have been centrally designed at the beginning of the season and thus rarely take into account the actual development of the pest situation and its agro-ecological context during the growing season. Such programmes suggest to farmers that pesticides are an essential part of modern agriculture and often draw them into unnecessarily high spending on pesticides.

A special case is the supply of pesticides for migratory locust control, which often results in flooding local markets with cheap pesticides and frustrates efforts to introduce IPM. Donations for migratory locust control often remain partially unused for a variety of reasons (arrives too late, volumes requested were too large, etc.). Left-overs either become obsolete or are used for different purposes, often free of charge or at very low costs. There have also been examples of countries that used locust threats to request pesticides in order to stock up free pesticides for regular agricultural purposes[10].

Moreover, the economic justification for large-scale spraying campaigns to prevent or control locust outbreaks is subject to debate[11]. Countries prone to locust outbreaks are investing in locust control, often with external financial assistance. They are paying a high price to "insure" against the possibility of severe economic impacts associated with a locust plague. According to an FAO study[12], this price appears to be uneconomic in 80%-90% of likely circumstances. Requests for assistance in locust control are generally accompanied by standard statements about food security and impacts on the rural poor. Although such risks exist, it should be noted that there is a large variation in actual circumstances and that serious impacts are rare. Under most circumstances control efforts are questionable in terms of providing a cost-effective response to food security risks. While the interests of the rural sector are often cited as an important factor in the justification of spending on locust control, the benefits of control efforts are generally enjoyed largely outside the rural sector.

Emotional and political arguments for chemical control often seem to prevail over economic arguments. Politicians want to be seen acting when swarms of locust appear on the TV news. Donors want to be seen responding to politicians. In addition, chemical control is often promoted by parties that have an interest in locust campaigns, ranging from the pesticide industry to national plant protection services, which often depend on locust campaigns for their vehicles and allowances.

3.2   Lessons learned about hazards of pesticide use

This section provides an overview of the lessons learned about the side-effects of pesticide use in developing countries. It is a brief introduction which focuses on some salient points that are thought to be relevant and of practical use to staff involved in appraisal of requests involving pesticides. For more detailed information on pesticide use in developing countries, reference is made to the list of recommended literature in Appendix 2.

• **Pesticide use has negative implications for public health and the environment**
Worldwide, twenty to thirty billion dollars worth of pesticides are released into the environment each year. This has commonly led to widespread environmental pollution affecting fauna, flora and ecosystems. Excessive and injudicious use of pesticides has also contributed to diminished biodiversity.
Excessive use, inappropriate disposal practices, and leakage as a result of inappropriate transport and storage, have, at many locations, caused dangerous levels of soil and water contamination. People in developing countries often depend on a single water source. Contamination of these water sources has severe public health implications.

Besides the risk of acute poisoning when handling pesticides, farmers run the risk of gradual poisoning through repeated low-level exposure. Farmers and the public are further affected through gradually increasing contamination of the environment and through direct intake of pesticide residues with food and drinking water. Proven longer-term effects of gradual poisoning include: cancers, birthdefects, reproductive anomalies, skin problems, etc. The longer-term effects are still poorly understood and may be worse than expected so far. A new scientific debate is arising on the possible effects of first generation pesticides (organo-chlorine compounds) on fertility, the endocrine system and the auto-immune system[13]. It was 20-30 years after the use of these products peaked that this issue firmly appeared on the international agenda. The list of pesticides suspected of endocrine-disrupting properties is steadily growing and increasingly includes more modern pesticides. New scientific insights may emerge in ten years' time about today's pesticides.

A specific environmental problem is the huge stockpiles of obsolete pesticides that are found throughout the developing world. Pesticide donations have been a major factor in the accumulation of obsolete stocks[14]. Obsolete pesticides can no longer be used because their use has been banned or because they have deteriorated as a result of prolonged storage. FAO estimates the total quantity of obsolete pesticides in Africa at 15,000 to 20,000 tonnes. The situation in Asia and Latin America is less well documented but is expected to be worse. In Africa, most of the obsolete stocks are stored in sub-standard stores in urban areas and many containers are leaking. Presently, the recommended disposal method is incineration in a special hazardous waste incinerator. With the exception of a few newly industrialised countries, developing countries do not have such incineration facilities. This means that the waste needs to be exported to a hazardous waste incinerator in a developed country. The overall cost of export for incineration ranged from USD 3,000 to USD 4,000 per tonne for disposal operations conducted during the period 1992-1997.

• **Most of the pesticides commonly used by small-scale farmers are not safe**
An estimated 25 million farmers and agricultural workers in developing countries suffer pesticide poisoning each year[15]. The main cause is unnecessary occupational exposure to pesticides as a result of general overuse. Other contributing factors include easy access

to inappropriate pesticides (e.g. too hazardous because of high toxicity), lack of afford-able protective clothing, and general ignorance about pesticides and alternatives. Many farmers and plantation workers have adopted a fatalistic attitude and accept suffering poisoning symptoms as an inevitable consequence of spraying pesticides. They some-times even anticipate being ill for a day after spraying.

Pesticides listed in WHO Hazard Classes I, II and III[16] are hazardous and thus not safe. Normal use of these products by small-scale farmers involves risks. Because of the above factors, products of WHO Hazard Classes Ia and Ib (extremely and highly hazardous) and the higher range of Class II (moderately hazardous) are generally considered to be unsuitable for use by small-scale farmers[17].

However, in many countries the use of highly hazardous pesticides among small-scale farmers is still high. Instead of withdrawing these products, agro-chemical companies organise "safe use" training to reduce risks associated with them. Generally, such train-ing does not solve the problem because it does not reach all users, and the availability and cost of protective gear remain an obstacle[18]. In most cases it is also impossible for small-scale farmers to ensure safe storage of products and to adhere to prescribed proce-dures for the safe cleaning of application equipment and the safe disposal of left-overs. "Safe use" training for small-scale farmers is a misleading term because it incorrectly suggests that risks can be reduced to zero.

- **Government pesticide control schemes often do not work adequately and product stewardship by agro-chemical companies is less in developing countries than in Western Europe**

Generally speaking, farmers in developing countries are insufficiently protected against inappropriate pesticides. An FAO review of the implementation of the International Code of Conduct on the Distribution and Use of Pesticides in 1993 showed that most countries had enacted pesticide legislation and established a pesticide registration scheme. At the same time it was found that enforcement of such legislation was often insufficient and that the private sector had made no progress in addressing known and important shortcomings, such as inferior quality of marketed products, unsubstanti-ated claims in advertisements, and labels not conforming with official recommenda-tions[19].

Many examples illustrate that in practice large multinational companies apply different standards in developing countries. There is often a big gap between published policy statements on company ethics and actual practices of local company representatives, subsidiaries or sales agents in the field. Direct and indirect incentives provided by com-panies deflect rational decision-making regarding pesticide requirements. Pesticide reg-istration authorities, decision-makers on pesticide requirements, research stations, plant protection and extension staff often receive funding, commissions, fees or other incentives from pesticide companies.

3.3   Constraints of earlier IPM programmes and the development of participatory IPM

From the 1970s onwards there have been many aid projects to help introduce IPM in developing countries. Successes had been documented in tropical plantations and estate farms, but projects aimed at small-scale farmers, who are the main agricultural producers in developing countries, often had limited impact or not at all. The common design for IPM programmes was that a research station would develop an IPM solution to a specific problem and that extension staff would then instruct farmers on how to apply the recommended intervention. This approach did not work well. Somehow it appeared difficult to change farming practices in the longer term. Many people were therefore tempted to conclude that IPM was too complicated for small-scale farmers and that more research was required to develop better and simpler IPM packages. Another important factor was that agricultural policies and services strongly favoured chemical control as a mainstream approach to crop protection.

Experience of participatory IPM pilot activities demonstrated that neither the complexity of IPM nor the availability of scientifically developed IPM strategies, was the real bottleneck to large-scale implementation. Instead, the following factors were identified as common impediments:
- Aggressive marketing of pesticides as a result of increased competition between companies for their share of the growing local market;
- government policies and development aid directly or indirectly promoting the use of pesticides and thereby discouraging IPM, and absence of a supportive institutional environment for IPM;
- tendencies towards centralised decision making regarding pest control strategies. Central instructions to use pesticides invariably caused overuse of pesticides and frustrated IPM, while centrally designed IPM strategies were based on standard situations and did not take into account the wide variety of specific circumstances in farmers' fields;
- extension schemes with a strong top-down nature which are merely ordering farmers to follow instructions and tend to focus on male farmers;
- efforts to introduce IPM that focus on pursuing ideal IPM packages and remain research-driven instead of farmer-driven. Barely existing link between researchers and farmers, while link between researchers and extension staff were weak. Researchers got effectively no direct feedback from farmers and only little feedback from extension staff.

Farmers would stop practising IPM as soon as it no longer made sense to them, for instance because: the situation in their fields was different from what had been explained to them; unexpected problems occurred; pesticide sales agents convinced them that they should use their products instead. The lesson learned was that there is little point in trying to introduce IPM if farmers do not understand the underlying agro-ecological principles and if they have not learned to adapt IPM to their specific situation and needs. Participatory IPM provides an adequate response to these limitations.

Farmer Field School participants undertaking agro-eco system analysis, Cambodia (photo: Robert Nugent)

Participatory IPM combines a technical plant protection component with a social component and uses Non-Formal Education training methods. Farmers are no longer regarded as recipients of extension instructions, but rather as human resources capable of attaining insight into the agro-ecosystem of their crop and to take well-informed and sound crop management decisions by themselves. These capacities are developed in season-long Farmer Field Schools (FFS) on IPM. FFS are conducted by facilitators who have been trained during season-long, field-based Training-of-Trainer (TOT) courses (see Box 3).

Box 3:

What is a Farmer Field School and a Training of Trainers?

In a Farmer Field School (FFS) on IPM, farmers meet weekly during a full cropping season to conduct experiments and to monitor and discuss crop management interventions. The four key principles of FFS training courses are: (1) Grow a healthy crop; (2) Observe field weekly; (3) Conserve natural enemies; (4) Farmers understand ecology as experts in their own fields.

During a FFS, farmers learn field observation methods. Weekly observations compare IPM plots with plots managed under calendar-based spraying schemes or conventional practise. Plants are sampled and carefully observed while pest and natural enemy population sizes are monitored and recorded. Groups depict the situation in their field in drawings and present their 'Agro-Ecosystem Analysis' for plenary discussion. The participating farmers then decide what crop management practices will be applied and closely monitor the impact. Conservation and utilisation of local natural enemies and other beneficial organisms play an important role in the control of insect pests. Participants also look at other pests and at nutrient and water management. Pesticides (selective and with a low toxicity) are applied only after field observations have shown that they would supplement natural mortality and non-chemical management methods.

At least 30% of the FFS time is spent in the field. There are no standard recommendations or packages of technology offered. In the FFS, farmers collect data in their experiment plots and decide on interventions based on their findings. Experiments are conducted to demonstrate interaction between pest populations, natural enemy populations and pesticide applications. Other experiments demonstrate the relationship between initial visual plant damage, the crop's physiological capacity to recover from damage, and eventual yield losses. Farmers find out for themselves that limited damage does not usually reduce yields, and that spraying against several pests increases both production costs and the risk of further pest outbreaks. They also test extension recommendations. Through learning by doing, they gradually gain the confidence to make independent decisions on crop management. FFS uses visualisation techniques (drawing, seasonal diagrams, theatre, etc.) and much attention is paid to group dynamics.

Full-time season-long field-based Training of Trainer (TOT) courses are organised to prepare extension staff to conduct FFS training. During the TOT they carry out comparative experiments, and grow and monitor the target crops to learn about the problems that farmers face throughout a cropping season. As part of the TOT, and parallel to their own training, they also conduct their first FFS under guidance of experienced master trainers.

Gender awareness is built in at curriculum development stage and in selection strategies for facilitators and farmers.

Participatory IPM was developed largely under the Intercountry Programme on IPM in rice in Asia, which started in 1980 with technical assistance from FAO. There had been some successful small-scale NGO projects which used participatory non-formal education methods to train farmers in IPM. The challenge of the Intercountry Programme was to further develop this farmer-centred approach and to integrate it into government policies and extension programmes with the aim of reaching large numbers of farmers.

By 1996, the Intercountry Programme encompassed 12 Asian countries which had established national IPM programmes. Initially the programme concentrated on rice, but at the request of governments and farming communities it expanded to other crops, such as cotton, cabbage, tomatoes, French beans, soybeans. By 1997, over 1.5 million Asian farmers in over 60,000 communities had been trained through FFS and achieved massive savings on pesticides while increasing yields through better crop management. Government investment in IPM training began to outweigh earlier donor investments. Participatory IPM became increasingly recognised as a new standard for IPM programmes. The number of

African countries starting IPM programmes with FFS training is rapidly growing. Box 4 provides examples of participatory IPM programmes and projects ongoing in 1997/1998.

FAO has conducted detailed cost-benefit analyses of the farming practices of small-scale farmers involved in participatory IPM programmes. In over 20 countries in Asia and Africa, such analyses invariably demonstrated that IPM-educated farmers can grow rice, cotton or vegetables with significantly less pesticides while maintaining or increasing production levels and improving profits.

3.4   Specific lessons learned regarding participatory IPM

Important lessons learned from the Intercountry Programme and other participatory IPM initiatives are:

• **Farmers can become experts in agro-ecosystem analysis**
Tens of thousands of Farmer Field Schools in Asia and Africa have demonstrated that small-scale farmers can become active experts in agro-ecosystem analysis and can take well-informed crop management decisions based on their own observations and assessments. Their knowledge is supplemented by findings of field experiments they design and conduct themselves to find solutions to specific problems. Farmers, both men and women, acquire skills and create knowledge that puts them in control of farming technology.

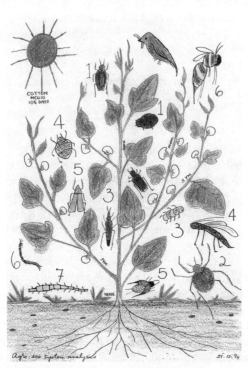

FFS-educated farmers have learned how experiments can help them solve problems. After the FFS, they generally continue to conduct field experiments on their own initiative to learn more about the agro-ecosystem and their cropping practices.

FFS-educated farmers can also become capable trainers. From each FFS, some farmers generally emerge who have a high commitment and good process skills. In Indonesia, by early 1998, 16,000 of such FFS educated farmers had received additional training to train other farmers. Generally they receive national programme support to conduct one FFS and thereafter play a role in organising farmers around follow-

Agro-eco system analysis made by farmer in Farmer Field School

up activities. The quality of FFS conducted by farmer trainers is equal to those conducted by TOT-trained extension staff.

Besides these national programme supported Field Schools, there are many informal and direct actions that spread IPM. FFS-educated farmers often demonstrate the effects of IPM to farmers who did not participate in FFS. They are either asked to do so or take the initiative themselves. Farmers realise that the effect of IPM is greatest if it is introduced on a broader scale. They also find the issue interesting and like to share it with other farmers from their villages. Case studies in Indonesia have documented how religious or cultural tenets also motivate FFS-educated farmers to spread IPM among friends and neighbours.

> *"I am not using pesticides anymore. Evidently pesticides increase your cost and don't increase your yields. My friends and I can now identify pests and natural enemies. When we observe our fields we can determine our potential losses and whether we are going to lose any money. It is clear, if we spray we will kill natural enemies. This is a cost, it will lead to there being more pests."* (Madkur, IPM farmer, Lampung, Indonesia)

> *"There are lots of natural enemies in my fields now. Of course there are pests, but not many. I have profited from IPM, my costs are down, yields are good, and my ecosystem is safe."* (Paiman, IPM farmer, Lamoung, Indonesia)

> *"In truth many farmers will just follow what other farmers do. You do not need to force them if they see you doing something that saves them money and is effective. IPM lowers your costs so your income increases. IPM is a healthier approach for everyone, and has become very important to us."* (Rowi, IPM farmer, Ngantang, East Java, Indonesia)

- **Participatory IPM contributes to community development**

After FFS education, farmers generally become more confident about their business, better organised and more responsive to community needs. FFS-educated farmers often take the initiative to set up IPM-based farmer organisations, or reorganise or strengthen existing farmer groups to serve as a forum for IPM issues. These organisations or groups play an important role in the further development and spread of IPM, and in obtaining support for IPM from local government and extension services.

Various examples demonstrate that farmers who believe in IPM can effect policy change. These include cooperatives that changed their attitude to pesticides and district-level pest control policies that were changed to support local IPM initiatives. IPM/FFS is increasingly recognised as an entry point to farmer empowerment which helps them take better control of their own economic and social well-being.

> *"For the last three years farmers have not accepted pesticides as part of the credit package of the Village Cooperative Unit. When they have been forced to take pesticides the farmers have returned them to the cooperative."* (Haji Mustofa, IPM alumni, West Lombok, Indonesia)

"Because of our study we have succeeded in involving all farmers in controlling rats, IPM and non-IPM farmers alike." (Pak Hani, IPM farmer, Kaligondang, East Java, Indonesia)

"Now you find that we have a farmers movement. Without waiting to be told to do so, farmers are organising… Now we village officials just follow along." (Pak Soepono, village head, Cilapar, Central Java, Indonesia)

- **Farmer Field Schools have a positive impact on the position of women**

Within agriculture, plant protection is still often seen as a male domain. Plant protection extension is therefore generally biased towards men. Women's knowledge about pesticides is limited, yet they are regularly in contact with pesticides, both in the field and outside the field when processing crops.

FFS offers good openings for the empowerment of women. Besides saving money and reducing health hazards, they learn new agricultural skills and become aware of hazards associated with pesticide use. They discover that they can design and conduct problem-solving experiments. Their confidence in local knowledge is strengthened. They make charts despite not being able to write and present their findings in drawings to an audience which includes male farmers. They discover that (male) extension staff listen to them instead of just telling them what to do.
Female FFS participants often view FFS as an ideal opportunity to learn skills which change them from implementers into decisionmakers. Given the chance of participating in FFS, female farmers are often very dedicated and show better attendance rates than male farmers. After their FFS they often take the initiative to inform or train other women in IPM.

Experience in major Asian programmes indicates that explicit strategies are required to achieve female/male target ratios in FFS. Elements of such strategies are: gender-labour analyses of farm work as part of selection procedures for FFS participants; policies to increase the number of female FFS facilitators; improved gender awareness of trainers by incorporating gender issues more explicitly in their Training-of-Trainer curricula; or just straightforward all-women-FFS. FFS should be planned on the basis of the results of the gender-labour analysis. The hours of FFS activities should match women's work schedules and the location should be convenient to women. Studies in Indonesia have shown that low participation of women is usually not caused by cultural restrictions, but rather by low gender awareness among trainers and government officials. In 1997, the ratio of female participants in Indonesia was about 20% (up from 3% at the beginning of the programme). In Vietnam and the Philippines ratios were around 30%. Efforts are being made to further increase this ratio.

"My husband's income is not enough for us to live on. The area I farm is only 750 m2 so I have been interested in finding a way to farm more effectively so that my yields might be as high as possible. Before I took part in the Field School I farmed the way my parents had taught me. I used Urea and TSP and applied Diazinon two or three times a season. I usually made applica-

tions to control brown plant hoppers. Later, when rice seeds bugs appeared, I sprayed again. Finally, just before harvest, I would put on a final application to insure against damage. My yields averaged between 200 and 300 kg. After attending the Field School I changed my approach to farming. I learned that by applying pesticides I was increasing my costs as well as increasing my risks. Both pests and natural enemies are killed by pesticides. If I don't spray, the natural enemies do my pest control work for me. The field school also helped me to learn about balanced fertilisation and planting distances. I first started to apply IPM principles without telling anybody. My yield for the first season went up to 350 kg. Since then I have averaged 400 kg. Since then I have not been quiet about IPM. I meet with women's groups and tell them about IPM principles and the dangers of pesticides. I tell the farmers in the fields around me to watch what I am doing and learn from me. Other farmers have been watching me and are following me. No one near me is applying any pesticides." (Ms. Romini, IPM farmer, Purbalingga, Central Java, Indonesia)

"We know you can grow long beans without pesticides, we proved this. In the next village over, a farmer sprayed his long beans. Hey, there wasn't one women from here that bought any of his long beans." (Ms. Srimulat, IPM farmer, Brecek, Central Java, Indonesia)

"We will establish a capital fund to support more group activities and I am pushing the Village Head to fund a Field School for our husbands who, for the most part haven't attended a Field School." (Ms. Sutiyem, IPM farmer, Brecek, Central Java, Indonesia)

Farmer Field School, Philippines (photo: FAO)

• **Participatory IPM requires extension staff to adopt a new role**

Participatory extension methods are essential to IPM. Conducting Farmer Field Schools requires extension staff to adopt a new role which is quite different from the role of staff in typical top-down extension systems. Instead of being intermediaries who deliver pre-formulated "one-size-fits-all" extension messages and instructions to farmers, extension staff need to become facilitators. Instead of being passive recipients of instructions, farmers become problem-solving actors who monitor the situation in their fields through observation and experimentation. The process becomes more important than the technical recommendations that result from it.

FFS puts farmers first and changes the relationship between farmers and extension staff. During FFS much time is spent in farmers' fields and in discussions emerging from observations in their fields. Learning becomes a dynamic process of interaction during which farmers actively contribute to finding solutions to their problems. Farmers no longer wait for extension staff to tell them what to do, but find out for themselves what to do and approach extension staff if they have specific questions.

The question that arises is how realistic it is to change extension staff, who are biased towards pesticide use as the obvious mode of crop protection, are used to lecturing passive target groups, derive their knowledge from lectures by Subject Matter Specialists, and may have annual targets in terms of the number of farmers that need to be reached (there is always a shortage of extension staff) or quantities of inputs that need to be distributed. Training-of-Trainer (TOT) courses are designed to prepare extension staff for their new role and include non-formal education methodologies and facilitation skills. Hundreds of TOT courses in more than 20 countries have demonstrated that it is possible for instructors to become facilitators and to remove biases towards pesticide use. Extension staff are often people who have chosen the profession because they like to work with farmers and believe that they can achieve something. Just as often they become frustrated for of various reasons. Conducting FFS often appears to bring back something of their earlier motivation and to restore job satisfaction.

In the early stages of IPM programmes most facilitators were government extension staff, mainly plant protection extension staff, but also general agricultural extension staff, and NGO extension staff. In the later stages use is also made of farmer trainers. These are farmers who passed through FFS and demonstrated an ability to educate and motivate others, and received some additional training to prepare them as trainers.

Post-FFS follow-up activities are needed. In Indonesia, IPM Farmer Forums are being organised at sub-district level. These forums support IPM farmers in their efforts to develop farmer-led IPM activities and create their own knowledge systems. Plans developed in the context of these forums are often funded by local government or form a basis for the allocation of national Programme activities.

- **Lack of scientific information on IPM techniques is generally not an obstacle to IPM implementation**

For a long time, much of the investment in IPM has gone into research programmes because the availability of scientific information was regarded as an important obstacle to effective IPM implementation. However, for a range of crops, experience in initiating participatory IPM clearly indicated that scientific IPM information was not a factor impeding the commencement of IPM. Generally, there was sufficient information available, both among researchers and among farmers themselves, to provide a basis for initiating an IPM programme. It just needed to be brought together and integrated into curricula for TOT and FFS which served as a basis for further problem-solving experiments conducted by the farmers themselves (see also box 5).

Linkages between research programmes and farmers are generally very weak. Research

Training of Trainers Insect-zoo, Vietnam (photo: Paul ter Weel)

findings are rarely found to be directly relevant to farmers' field problems. Research stations often operate without direct feedback from farmers. A review study in Asia[20] identified the following reasons for the lack of linkages between research on the one hand and extension and farmers on the other:

- mutual distrust: researchers do not believe that farmers can do science; farmers and FFS facilitators think that researchers are too academic to be of any use;
- researchers are not interested in applied, on-farm research, they have a fundamental research agenda;

- organisational structures of ministries, with research and extension under separate (advisory/supervisory) hierarchies, lack mechanisms to link them together.
- lack of research information was often not the problem in the early participatory IPM programmes which focused on rice (nor was it for later programmes involving other crops).

As a result of these poor linkages, many findings of research institutes working on IPM remained unused. The experience with participatory IPM calls for a review of the role of research. Researchers can enhance IPM programmes if they are willing to operate in partnership with farmers, and if research agendas respond better to needs identified by farmers. For participatory IPM programmes, investment in extension is far more important than investment in institution-based research.

- **Ownership is a key to sustainability**

FFS-educated farmers feel that IPM has become something of themselves that gives them control and provides a reference when talking to extension staff. They gain confidence and become more assertive towards visiting pesticide salesmen. They would not readily abandon their IPM practices because they understand what they are doing and know the ecological and economic implications of pesticide use. It is this ownership of knowledge that proved to be the key to sustainability. The feeling of ownership is also an important factor that prompts FFS-educated farmers to create learning opportunities for other farmers.

Ownership is also very important at policy level. National IPM programmes function better in countries where there is high-level policy support. Such support derives from involvement. Experience in Asia and Africa has demonstrated the importance of involving high-level policy-makers in the development of IPM activities from an early stage. Many projects were even started with study tours for policy-makers to countries where the implementation of national IPM programmes was in full swing. After having seen participatory IPM at work, and having talked to their local colleagues and to farmers practising IPM, they often returned as strong advocates for the introduction of participatory IPM in their own countries. Early involvement of key senior government officials in all stages of project development is now considered a vital element in sustainability.

- **Participatory IPM is applicable in different agricultural production systems**

The need for the introduction of IPM is best understood in circumstances characterised by:
- decreasing yields as a result of crises in pest control;
- increasing production costs due to increasing use of pesticides;
- greater awareness about health and environmental problems related to intensive pesticide use.

This means that crops grown under intensive production systems such as rice, cotton, vegetables and plantation crops, are generally good targets for IPM.

The fact that the biggest impact is achieved on crops that have entered a crisis stage does not mean that participatory IPM would be of less interest to small-scale subsistence farmers who practise mixed cropping with low levels of agro-chemical inputs. Participatory IPM means far more than saving money on pesticides. Although insect pests provide the easiest starting point for FFS, there are many other problems that are being addressed in FFS. Participatory IPM is also about the control of weeds, pathogens and rodents, and about making more effective use of fertilisers, irrigation water and seeds. FFS is about an *approach*, not about a technical issue. The *approach* encourages farmers to identify and tackle their own problems. During the first series of FFS in Laos, rice farmers for example discovered that they could increase yields by over 20% by more effective use of the same amount of fertiliser. The FFS approach would also be useful for projects working on the introduction of Low External Input Sustainable Agriculture.

> *"Our group came up with the idea of studying the use of urea tablets and Super Phosphate because we wanted to see if there was any proof for what had been promoted by the extension officers."* (Haji Suwito, IPM farmer, Tejasari, Central Java, Indonesia)

> *"I will conduct a field study on fertiliser and a varietal trial so that I can learn more. What I learn I pass on to others. I am trying to influence the others in my farming group to use IPM. I want to see the day when no pesticides are applied in the rice fields in this hamparan."* (Paiman, IPM farmer, Lampung, Indonesia)

FFS pilot activities with small-scale farmers growing vegetables, maize, cowpea, groundnuts, etc. showed that these farmers were very interested in FFS training. In Ghana, a participatory IPM pilot project focused on farmers growing rice in irrigation schemes. The next year, extension staff and farmers took the initiative to develop a participatory IPM approach for cassava and cowpea. There are more examples of IPM starting with an intensified crop and moving on to a subsistence crop. In Indonesia farmers started with rice and now apply IPM to a broad range of crops.

Moreover, it generally appears that eagerness to learn about their crops is the primary motivation of farmers for wanting to participate in FFS. The usefulness of participatory IPM is certainly not restricted to farmers in intensified "green revolution" agricultural production systems. However, if there is a choice, there might be a preference to start with farmers in intensified production systems because the impact is more easily demonstrated. Box 4 provides examples of ongoing participatory IPM projects and demonstrates the variety of crops and production systems.

Box 4:

Examples of participatory IPM programmes and projects ongoing in 1997

Country	Crop	Starting year
Bangladesh	rice; egg plant	1989
Cambodia	rice; vegetables	1995
China	rice; cotton; vegetables	1991
India	rice; cotton; vegetables; oil seeds; pulses	1987
Indonesia	rice; vegetables; soybean; potato	1986
Malaysia	rice; vegetables; fruits	1993
Lao PDR	rice; vegetables	1996
Philippines	rice; vegetables; maize; coconut; mango	1986
Republic of Korea	rice; fruits (apple and citrus)	1994
Thailand	rice; vegetables	1995
Vietnam	rice; cotton; vegetables; soybean; tea; maize	1992
Burkina Faso	rice	1996
Cote d'Ivoire	rice	1996
Egypt	vegetables	1997
Ghana	rice; vegetables	1995
Kenya	coffee; tomatoes; cabbage, maize, beans	1995
Mali	rice	1996
Sudan	cotton, vegetables	1995
Tanzania	rice; cotton, maize	1998
Zimbabwe	cotton, soybean, vegetables, paprika, maize	1997
Honduras	maize, beans	1991
Nicaragua	maize, beans	1992

- **Intercountry cooperation and coordination enhances national IPM programmes**
  At project level, TOT courses have benefited from invited experienced master trainers from other countries. Relatively inexperienced trainers from countries with new programmes graetly benefited from participating in TOTs in countries with more advanced IPM programmes.

At policy level, exchange visits and regular regional meetings offer participants an opportunity to exchange experiences with colleagues from other countries and to discuss institutional reform (pesticide pricing policies; reform of extension services and research programmes; etc.) required for an effective consolidation of IPM programmes. It offers countries a chance to benefit from experience gained in other countries that are further ahead in the consolidation process.

3.5 Constraints of participatory IPM

Scarcity of longer-term impact data
Many of the participatory IPM activities are still relatively new. So far most impact studies have concentrated on comparing yields and profits of FFS-educated farmers with local farmers' practice. Such studies are important at the beginning of a national programme because they generate support and commitment from government, farmers, communities and extension services.

Only a few national programmes have reached the consolidation stage (see later) and the actual consolidation process is still under development. Finding the right modalities for creating a capacity to organise vast numbers of FFS and for follow-up after FFS is a dynamic process. Strategies are tested and adjusted all the time. At such a stage there is much quantitative information regarding the period immediately after FFS, but detailed quantitative information on the longer-term impact is still scarce. Despite this development-stage-related short-coming, there is agreat deal of qualitative information indicating that farmers continue with IPM, that attitudes change and that IPM is integrated into local agricultural policies. Qualitative information for instance also indicates that cooperatives remove pesticides from their input programmes and that local pesticide salesmen face rapidly dropping sales.

An important outstanding question is how the impact of participatory IPM should be measured. The overall impact is determined by many different indicators which vary from short-term to longer-term, from direct to indirect, from easy-to-quantify to difficult-to-quantify, and from farm-level to national-level, which makes it very difficult to obtain a complete picture. Examples of success indicators include: improved farm-level yields and profits after FFS training; spread of IPM after FFS training and impact at community level; overall reduction in pesticide use; improved occupational health and reduced loss of lives, working time and medical costs due to poisoning; decreased damage to the environment and water resources; indirect benefits from better organised farmers and new inputs into community development; decreased dependency on ineffective extension services; improved impact after reform of extension services and better returns on investment in extension; reform of research programmes leading to better agricultural returns on investment in research; longer-term food security; etc. To single out one, or a few, of these factors would give a distorted picture.

Because of the limited availability of quantitative information on the longer-term impact it is possible that some constraints have not yet become visible.

Constraints related to the policy environment
It takes time to carry out broad and sustainable policy changes to remove well-established institutional biases in favour of chemical control and to reform extension services to make these more supportive of an FFS approach. Often such changes are under way

but have not yet reached a critical momentum to adequately curb biases towards conventional chemical control. The result may be an inconsistent mixture of signals given by different government departments which each steer crop protection in different directions. Such inconsistencies may have an inhibiting effect on IPM programmes. A related factor is the stability and continuity of extension services. Unstable extension services with a high rate of official transfers of staff may have a large drop-out rate for IPM/FFS-trained extension staff.

Concerns related to the duration of TOT courses and FFS

A common concern about participatory IPM programmes is the duration of season-long TOT and FFS and the demands this would place on extension staff's time. Particularly in countries with an unfavourable extensionist/farmer ratio, these concerns may surface in the early stages of programme development. However, in the longer term the initial investment in time translates into sustainability of impact. Under more conventional extension systems, such as Training and Visit (T&V), extension staff continue to visit the same farmers for a longer period of time, while FFS-educated farmers manage more quickly on their own and require far less follow-up. Some follow-up is still needed, but because it enhances skills, understanding and community development it decreases dependence on extension services. Under FFS, extension becomes demand-driven (farmers ask extension services to advise them on specific issues), while conventional extension systems are predominantly supply-driven (extension services decide what they are going to tell the farmers). Demand-driven extension is more effective than supply-driven extension. Concerns about investment of staff time often disappear once the impact of FFS has been demonstrated. Heads of plant protection and extension services who have seen participatory IPM at work in another country tend to be less concerned about assigning their staff to take part in TOT and FFS.

Experienced farmers can become FFS facilitators and thus reduce the dependency on government staff. In Indonesia and Vietnam, more than half the FFS facilitators are experienced farmers who passed FFS, practised IPM and were then trained in programming and facilitating FFS. They receive national programme support to organise at least one FFS in their community. Thereafter they often play an important role in organising follow-up activities in their communities.

In cases where the impact of participatory IPM was less than expected, this mostly turned out to be attributable to attempts to shorten TOT or FFS, either by reducing the number of days or the number of activities. Such short-cuts dilute the quality of the training. Many experiments with variations on TOT and FFS clearly demonstrated that training needs to be regular (weekly) and season-long in order to achieve the desired impact. Monitoring the quality of TOT and FFS in expanding programmes is necessary.

Directly related to concerns about the duration of FFS are concerns about the costs. The costs of FFS vary and depend on the specific local situation. An important factor is whether the FFS is organised within commuting distance from the duty station of the extension staff involved. FFS conducted by extension staff and funded from project funds tend to be more expensive than those organised at a later stage by farmer-trainers with local community support. The general picture is that the costs of FFS are recovered by savings on pesticides during one or two seasons, while the horizon for savings continues far beyond that period into the future. In addition there are many indirect savings related to health, environment and food security. In Asia, a typical programme-funded FFS would cost USD 500 if conducted by extension staff and USD 200 if conducted by farmer trainers. In Africa, the costs of FFS conducted by extension staff may be higher because farming areas are often more extensive and extension staff may require temporary accommodation because commuting would take too much time. Early indications are that the average costs of FFS in Africa are in the order of USD 1000.

Farmers discussing the results of agro-eco system analysis in IPM Farmer Field School, Cambodia (photo: Robert Nugent)

guidelines for support for participatory IPM

This section provides suggestions for ways in which participatory IPM can be supported. It also discusses gender aspects and some of the pitfalls that embassy staff should be aware of.

4.1 Supporting participatory IPM projects and programmes to increase the number of farmers practising IPM

The main ways in which donors can support participatory IPM are:
- assistance with the establishment and development of national IPM programmes;
- assistance to relevant NGO initiatives;
- incorporation of participatory IPM components in ongoing or future projects or programmes that are not specific IPM projects, but address agriculture, rural development, community development, poverty alleviation, etc.

These possibilities are worked out in detail in the following sections. In addition, appendix 3 provides an overview of suggestions for activities that can be undertaken by NEDA staff to promote and support participatory IPM.

Establishment and development of national IPM programmes based on participatory IPM

There is no blueprint for IPM programmes. In each country, and for each cropping system, a unique IPM programme must evolve over a number of years as a result of practical dialogue among farmers, local officials, extension staff, researchers and policy-makers. However, based on experience in over 20 countries, the Global IPM Facility has identified the following four general stages in the development of national IPM programmes:

1. awareness building stage,
2. development stage,
3. capacity building stage,
4. consolidation stage.

These stages are further explained below. National IPM programmes and related donor support preferably should be planned on the basis of this four-stage model. Separate projects can be, and often are, funded for each stage. For the capacity-building stage, donor commitment is needed for the full

Holding harvested rice, Cambodia (photo: Robert Nugent)

period of at least 5 years recommended for this stage. Interruption of capacity-building projects because of discontinued funding tends to cause major setbacks. Ideally, a funding commitment should be made for a 7-8 year period, comprising both the development and capacity-building stage.

Awareness building stage

Activities in the awareness-building stage generate human and financial commitments to the testing of IPM by farming communities. Awareness-building activities have included study tours, workshops, field visits, farmers' presentations and brief pesticide policy studies. These activities are specific and time-bound. Usually they are funded with grant aid and cost less than USD 100,000. Awareness building activities may involve policy-level government staff, extension staff, farmers, farmers' associations, NGOs, universities, journalists, etc.

Development stage

This stage starts once awareness has built up and a commitment is made to field-test IPM implementation. Pilot projects are the main activities at this stage. Generally, these involve:
- a detailed survey of locally and internationally available IPM information regarding the crop;
- identification of the main actors and potential resource persons;
- identification of a trial site and target group;
- development of a gender strategy to ensure optimum participation of women;
- a curriculum development workshop for the TOT and FFS (see Box 5);
- development of training materials;
- a first TOT and a first series of FFS;
- evaluation and documentation of the results;
- more advanced pesticide policy studies;
- preparation of an outline for a national IPM programme.

Funding for IPM development activities, especially pilot field projects, has come from multilateral and bilateral agencies and national implementing authorities. Pilot projects take 18-30 months. Costs vary from case to case. Based on past bilateral or multilateral projects, an indicative amount for a pilot project would be USD 350,000.

Capacity-building stage

Activities in the capacity-building stage concentrate on preparing a national cadre of full time IPM FFS facilitators. This involves:
- design of a gender strategy to enhance involvement of women both at beneficiaries level and programme implementation and management level;
- identification of potential trainers and selection of TOT participants. These could be plant protection extension staff, other agricultural extension staff, NGO extension staff or experienced FFS-educated farmers;
- organisation of TOT courses;

- design and implementation of the first phase of a large FFS training programme;
- ongoing monitoring and evaluation of the impact of training;
- facilitating structural involvement of relevant research institutes, extension services and NGOs;
- further strengthening of policies conducive to IPM.

Capacity-building projects have been supported by national governments and (largely bilateral) grant donors. A typical capacity-building project would take 5 years. Costs vary from case to case. Based on past bilateral or multilateral projects, an indicative amount for a capacity-building project would be USD 3 to 4 million. For such an amount, the output would include several hundred FFS facilitators and several thousand villages in which FFS training has been provided. Roughly two-thirds of the budget would be spent on training and one-third on management and coordination.

Consolidation stage

During the consolidation phase large numbers of farmers will be trained and the programme will spread out to reach as many communities as possible. It includes follow-up activities to encourage and consolidate initiatives of FFS-educated farmers who organise themselves in order to continue experimenting and to spread IPM in their communities. This phase can start once a cadre of highly qualified and experienced full-time FFS facilitators exists and other prerequisites have been fulfilled, including supportive policies and a system of monitoring and evaluation to ensure the maintenance and quality of the

IPM schools Thailand with schoolkids (photo: FAO)

programme. This phase lasts a number of years and is usually supported by composite funding comprising both national and aid sources. A component of grant-financed technical assistance is required to guarantee the quality of training, assess the impact and adjust implementation in response to external changes in the field during implementation. Consolidation-phase IPM activities can either be stand-alone projects or can be integrated in wider sectoral investment projects.

Assistance to relevant NGO initiatives

In many cases, NGOs play a catalytic role in raising public awareness about the hazards of pesticide use and the need to change to IPM-based production systems. In several cases NGOs contributed to the development of participatory approaches and the setting up of participatory IPM field activities. Generally, NGO projects on participatory IPM would be similar to projects in the awareness-building and development stages as described above. Besides specific participatory IPM projects, there may be possibilities to build IPM components into other NGO projects. Some examples are given in the next section.

Incorporation of participatory IPM components in ongoing or future projects that are not specific IPM projects

It is worth investigating whether there are possibilities to incorporate participatory IPM components in ongoing or future projects or programmes on agriculture, rural development, community development, poverty alleviation, etc. Most projects that affect farming practices have such a potential. These include irrigation projects, farming systems projects, integrated agricultural development projects, agricultural credit schemes involving input supply, projects related to specific commodity crops, agricultural sector investment programmes financed through development banks, etc. Other projects that may have a potential for participatory IPM include environmental education projects, women in development projects, integrated rural development projects, etc.

Small investments in participatory IPM could enhance the impact of large and capital-intensive projects. A good example is irrigation projects. Such projects tend to focus on engineering and management aspects. However, the ultimate objective of increasing food production under an intensified system requires more than effective water supply. The benefits of good engineering and sustainable management can be, and often are, easily lost if crop production practices are not sustainable. Large-scale irrigation schemes in particular, tend to have production systems based on conventional "green revolution" input packages. To avoid unnecessarily high production costs and long-term destabilisation of production, all irrigation projects should have an IPM component. An interesting example comes from Ghana, where the irrigation authority enrolled its extension staff in TOT courses under the national IPM programme because it was realised that without IPM much of the potential gains of the irrigation schemes would be lost due to ineffective and unsustainable production. Vegetable production in these schemes was already suffering severely from spiralling pesticide use.

For new agricultural projects, participatory IPM should be built in at the design stage. The introduction of IPM has often been a reaction to a crisis situation after pesticide use had destabilised production. The new challenge is to move from crisis control to crisis prevention and sustainable intensification.

Besides incorporating participatory IPM components into Dutch funded projects, consideration could also be given to funding small satellite projects on participatory IPM that supplement larger and broader projects of the above categories funded by other donors (bilateral or multilateral).

4.2   Quality of training and post-FFS follow-up

When programmes reach the consolidation stage it becomes particularly important to pay attention to:
- quality assurance schemes for TOT and FFS training;
- follow-up activities for FFS-educated farmers;
- integration of gender issues in programmes and policies;
- long-term impact studies.

(See also some of the points raised in section 3.5 on constraints.)

Training courses may get diluted or simplified over time, which quickly tends to affect the impact. TOT and FFS quality monitoring and TOT refresher courses should therefore be components of the consolidation process. Generally speaking, such components will require external technical assistance.

The objective of participatory IPM programmes is that farming communities switch to IPM-based production. This is not automatically achieved by turning out large numbers of FFS-educated farmers. Follow-up activities are needed to help these farmers proceed with IPM after FFS. Experience in countries in the consolidation stage shows that many FFS-educated farmers take the initiative to continue the process of learning through experimentation, to spread the introduction of IPM through their community and to imbed IPM in local policies and agricultural services. National IPM programmes should actively monitor and facilitate this local-level consolidation process.

Specific activities to enhance the involvement of women are described in sections 3.4 and 4.4.
The need for long-term impact studies is explained in section 3.5.

Since several countries are going through the consolidation process and new lessons are still being learned, it is important not to operate in isolation but to exchange experience with other countries that are at the same stage and to take advantage of experience gained in these countries. The Asian Intercountry Programme and the Global IPM Facility play an important role in facilitating such information exchange.

4.3	## Policy reform

Agricultural practices are influenced by a range of government policies and regulations. These may enhance or inhibit IPM (see also 3.1). Policy adjustment may be desirable to provide better possibilities and incentives for sustainable agriculture. Donors and aid agencies can play a role in encouraging policy reform. Some suggestions:

• **Support studies on the impact of pesticide use on human health and the environment**

Funding of short studies on the impact of pesticide use on human health or the environment can be very useful. The results of such studies tend to be picked up by the press and advocacy groups and may have an impact on policies. Donors and aid agencies are often also in a position to raise the matter of pesticide use in environmental action plans and public health policies.

• **Support pesticide policy studies**

Pesticide policy analysis, if conducted thoroughly, helps to present a clear picture of the costs and benefits of pesticide use, and exposes inconsistencies in government policies. The real costs of pesticide use are often diffused by indirect subsidies and hidden costs to governments and farmers. These cost aspects are often not fully understood and therefore tend to be ignored, which may significantly distort the comparative production costs of IPM-based production systems versus chemical-control-based systems. Pesticide policy studies often serve as eye-openers for government policy-makers and budget-holders regarding government wastage of funds on pesticides. Examples of indirect subsidies and hidden costs are provided in section 3.1. Guidelines for pesticide policy studies have been published by the Pesticide Policy Project of the University of Hannover[21].

• **Encourage the elimination of unnecessary pesticide use**

Promote a critical case-by-case review of all pesticide use to determine what is useful/necessary and what is not. The aim of such a review would be to identify and eliminate excessive use and any other use which is not necessary, because it poses to many hazards to health and the environment or is not justifiable in economic terms. This also applies to pesticide use in emergency campaigns to control locusts or other migratory pests. Factors to be reviewed include: actual necessity for chemical control; type of pesticides and associated hazards; quantity of pesticides and frequency of application; costs and benefits of pesticide use, including costs of any negative impact on health and the environment; etc.

Encourage replacement of conventional pesticides by newer products such as bio-pesticides or other non-toxic control chemicals if any control agents are required as last-resort external inputs.

- **Support projects that assist governments in strengthening their pesticide control capacity**

Enforcement of good pesticide legislation can help avoid inappropriate use of pesticides and associated health and environmental problems. Where relevant, governments could be supported in strengthening their capacity to control the importation, distribution and use of pesticides within the framework of the FAO International Code of Conduct on the Distribution and Use of Pesticides. Such support could include technical and financial assistance, such as: drafting or reviewing pesticide legislation; training of inspectors to monitor compliance with pesticide legislation; upgrading of existing laboratories to enable pesticide quality control analysis for the verification of pesticide specifications; etc.

- **Be critical and exercise restraint regarding requests for pesticide donations**

In general, it is the Netherlands Developmeny Cooperation policy to discourage the supply of pesticides under aid programmes. Even requests for emergency operations and locust control should be treated with restraint. If, for any exceptional reason, pesticides are nevertheless supplied, this should be done in line with the OECD Guidelines for Pest and Pesticide Management and in coordination with other donors. The relevant section of the OECD Guidelines on "ensuring good practices when providing pesticides under aid programmes" is attached (Appendix 4). It includes a checklist for the appraisal of requests for pesticides. Requests for pesticides could be used to begin a discussion with government on the necessity of pesticide use.

4.4   Reform of agricultural extension and research

Participatory IPM programmes require a new role for plant protection extension services, which need to recognise the role of farmers in designing IPM strategies suited to the local situation (see 3.4). This has implications for the training and extension staff and the way their time is spent. Reform is therefore required at policy level where it is determined what role and budget extension services are allocated within the wider agricultural framework. Within budgets for agricultural production it may be possible to shift funds from input supply (procurement of pesticides, or pesticide subsidies) to strengthening extension services, as has already been done in several countries. Pilot projects help to demonstrate the soundness of such shifts in funding.

Further research is no longer regarded as a prerequisite for commencing field implementation of IPM (see also 3.4). Box 5 gives an impression of how a training curriculum is developed on the basis of existing information.

Box 5:

> **Getting started: Steps in designing a curriculum for TOT and FFS**
>
> *The process of designing a curriculum commonly involves the following steps:*
>
> 1. *A detailed survey to describe and evaluate local cropping practices (if possible in a gender-specific way), pests and diseases, IPM based practices, availability of relevant national expertise, etc. Collection of information/recommendations from ongoing participatory IPM projects in other countries and relevant national and international research/studies, etc.*
>
> 2. *A preparatory workshop to review and discuss the information collected, to identify the potential of various IPM techniques and to design a programme of experiments for farmer field validation of potential techniques. Participants should include selected farmers, relevant extension staff and researchers, invited experienced TOT trainers from a country which has already conducted a TOT on the crop concerned, and a gender specialist.*
>
> 3. *Selected farmers conduct the above experiments in their fields in close collaboration with extension staff and researchers.*
>
> 4. *A curriculum development workshop to draw up curricula for TOT and FFS. Such workshops may take up to two weeks and involve the same people as the preparatory workshop.*

Participatory IPM requires a different role for research institutes. It should be acknowledged that FFS-trained farmers can conduct problem-solving experiments and play an active role in research. Researchers should make explicit efforts to create partnerships with farmers and extension staff based on adequate, two-way interaction. Research agendas should be set in consultation with farmers and be responsive to needs identified by farmers. Research should be more site specific and involve farmers.

An appropriate balance needs to be found between investment in institution-based research and investment in extension programmes which enhance the utilisation of research results and at the same time facilitate farmer research. In many cases this would mean a shift of emphasis from the former to the latter.

4.5 Gender issues

As in all agriculture projects, special attention should be given to gender issues. The active involvement of women in agriculture, including IPM, is often still not sufficiently acknowledged in agricultural policies, programmes and projects. Women often have less access to agricultural knowledge than men due to biases favouring male farmers' participation in extension programmes. Participatory IPM programmes should, in their design, provide equal opportunities to women. This should be reflected in the ratio of female participants in TOT courses and FFS, as well as in the degree of involvement of women in project planning and management.

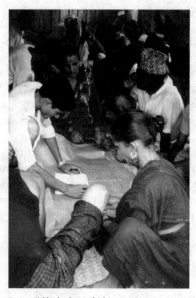

Farmer Field School, Nepal (photo: FAO)

Gender analyses should be made at all stages of programme development in order to secure optimum involvement of women and to improve the quality of the programme. Some specific recommendations in this respect are: selection of FFS participants on the basis of a labour-gender analysis (reliance on traditionally thinking government officials and local village leaders for the selection of FFS participants may result in a bias towards male farmers); involve a gender specialist in the curriculum development process for TOT and FFS; integrate gender awareness training into TOT courses; involve a national research institute specialising in gender issues in national IPM Programmes to monitor the programme and, at all stages, make recommendations on how the involvement of women can be improved; ensure that gender issues are specifically addressed in evaluation studies.

4.6 Pitfalls

There are a few pitfalls that embassy and project staff may come across and should be aware of.

• **Be aware of the ubiquitous bias towards pesticide use**

For decades it has been engraved in people's minds that pesticide use is the modern way forward to sustainable crop production. This conception has taken firm root in the majority of actors in agricultural development, ranging from farmers and extension staff to senior government staff. Such biases are not easily overcome and may result in project proposals that on the one hand aim at pesticide reduction and on the other reconfirm pesticide use as the prime mode of pest control. Examples include projects to reduce pesticide use by improving the effectiveness of application techniques or by training in the judicious and effective use of pesticides. One should always first investigate whether pesticide use is being taken for granted before considering projects aimed at pesticide reduction.

• **Be aware of stakeholders with an interests in maintaining the status quo in pesticide use and/or centrally designed crop protection strategies**

Large-scale implementation of participatory IPM means a shift from input supply to human resource development and from research to extension. Several stakeholders in plant protection may be affected by such changes and attempt to protect their interests.

One of the main stakeholders is the agro-chemical industry. In many cases, the significant pesticide reductions achieved through participatory IPM exposed the agro-chemical sector as a promoter of unnecessary pesticide use, and thereby added to the negative image the sector has been coping with as a result of a steady stream of publications about health and environmental problems caused by pesticides. The agro-chemical industry responded by declaring its commitment to IPM in formal policy statements and brochures. There is a genuine recognition that the market increasingly wants IPM and that in the long term the agro-chemical sector will have to adapt to this development. However, in the short term, companies have substantial interests in maintaining their market share for conventional pesticides. Despite industry claims that pesticides are becoming "greener" or "safer", conventional pesticides continue to account for almost the entire turnover of synthetic pest control products in developing countries.

Industry brochures on IPM make use of participatory IPM jargon, but at the same time promote industry's own interpretation of IPM which presumes that pesticide use is an essential element of IPM and that minimising pesticide use is a misinterpretation of IPM[22]. Industry interpretations often tend towards intelligent pesticide management based on selective and safe use of pesticides, where relevant in combination with non-chemical methods. Intelligent pesticide management maintains pesticides as the prime approach to crop protection and is regarded as a tool to secure agro-chemical markets, particularly where these are threatened by successful IPM programmes. "IPM training" is being organised by industry to promote the continued use of pesticides. New pesticides or other pest control agents which are more selective and/or less toxic are assigned IPM qualifications. Even older products known to affect the environment are sometimes presented as having such qualifications. IPM terminology is increasingly being abused in the marketing of pesticides.

In the late 1990s, the main focus of the agro-chemical industry is on genetically engineered crops. Billions of dollars are being invested in the development and introduction of such crops[23]. These are given new characteristics to enhance production or marketing. An important category are crops that are made resistant to specific herbicides to facilitate weed control. Typically this concerns herbicides produced by the company holding the patent on the seed. These crops will have a stabilising effect on the market for these herbicides. Industry claims that such crops will lead to an overall reduction in herbicide use. Critics contest this view and fear increasing use of herbicides and cross-over of herbicide resistance into wild plants. They regard genetically engineered crops as a risky and immature technology that is being rushed to the market in the scramble for market share among the companies involved. The debate on genetically engineered crops is complicated and has many aspects that can not be discussed in detail within the framework of this document.

However, from the viewpoint of participatory IPM, it could be noted that genetically engineered seed programmes are likely to keep agriculture corporate-driven, instead of farmer-driven. Farmers would remain dependent implementers of centrally designed input packages.

- **Beware of poor-quality IPM projects**

With the growing interest of donors in participatory IPM there are many parties who are trying to jump on the bandwagon to benefit from funding. Careful appraisal of funding requests is therefore important. Key elements that should not be missing are: intensive TOT courses for facilitators; intensive field-based FFS which take the local cropping system as a starting point for learning; policy reform; community-level follow-up.

Research projects, "safe-use" projects, or industry-designed intelligent pesticide management projects may misleadingly be packaged as participatory IPM projects. Such projects may even be dressed up as NGO projects. Industry sometimes presents itself as part of the NGO sector in order to get access to public fora and public funding[24].

- **Be critical about requests concerning projects for training in "safe use" of pesticides**

Safe-use training is not a solution to problems related to the use of highly toxic pesticides by small-scale farmers (see section 3.2). It is targeted at agricultural extension staff and reinforces the message that pesticides are the main mode of pest control. Sometimes it is also used to promote continued pesticide use in areas where such use may not be necessary from an IPM point of view, such as the use of insecticides for tropical rice.

This is not to say that training in hazard management is wrong. It is not, and it should be a standard element of industry stewardship programmes whose costs should be included in the price of pesticides. But it should not be used as a justification for continued supply of highly toxic pesticides to small-scale farmers, or as a promotional activity for the use of pesticides where such use is no longer justified. The agro-chemical industry has been seeking donor funding for their safe-use training projects. Donor support for safe-use projects should be regarded as an indirect form of subsidising pesticides because the costs of such training should normally be included in the price of pesticides. Support for agro-chemical industry projects for IPM or safe use would not be in line with the message of this Best Practice Document.

- **Replacement of pesticides by non-toxic pest control agents is not necessarily a solution**

Mere replacement of large-scale pesticide use by large scale use of biological control agents or other non-toxic products is generally not a complete solution to the negative impact of agricultural intensification. On the surface it may seem an attractive option, but in the final analysis it often will not eliminate the roots of the pest problem if it does not also put farmers in control of the agro-ecological process. Participatory IPM is explicitly broader than reducing pesticide use. It is about understanding the agro-ecosystem and enhancing its health through optimum use of (agri)cultural methods. Healthy agro-ecosystems have a better natural capacity to keep pests in check.

4.7   Coordination and collaboration

International coordination: the role of the Global IPM Facility

An increasing number of countries are interested in developing participatory IPM programmes and contact FAO or aid agencies for guidance and funding. At the same time many donors have given IPM a more prominent place in their development aid policies and are looking for good funding opportunities. A mechanism was needed to match demand with funding possibilities and to ensure high quality standards by connecting new initiatives to existing experience. The Global IPM Facility has been designed in response to these needs (see box 6).

Box 6

The Global IPM Facility

The Global IPM Facility was founded in 1995 by FAO, UNEP, UNDP and the World Bank and became operational in 1997. Its purpose is to help expand participatory IPM and to set standards for good participatory IPM. Its key activities are to:

1. *Create awareness and a favourable policy environment through study tours, exchange visits and briefings demonstrating the potential of IPM to farmers, technical leaders and policy-makers.*

2. *Help promote, design and facilitate funding for pilot activities to demonstrate the feasibility of a farmer-oriented approach. The Facility identifies and provides backstopping to skilled, experienced resource persons who can advise or guide these pilot activities. Such persons may assist with: selection of sites; formulation of criteria for selecting participants; preparation of training curricula; identification of principal trainers; enhancing involvement of policy-makers, researchers and academics; processing, analysis and documentation of results; etc.*

3. *Assist countries with successful pilot activities to move into a full-scale project phase. The Facility assists in identifying requirements and expertise for the design of projects. The emphasis is on strengthening IPM implementation through greater participation by national and local institutions, including NGOs and farming community organisations. The Facility brings governments into contact with potential donors and gives advice during submission and appraisal of the project document.*

4. *Monitor and evaluate participatory IPM projects and programmes, conduct case studies and develop standards for participatory IPM.*

5. *Help establish, strengthen and expand regional and national IPM programmes by providing linkages to other national IPM programmes and facilitating access to relevant models, experts, research findings and studies.*

6. Establish cooperative linkages with relevant officers, both technical and policy, within aid agencies, international agencies and NGOs and offer assistance in project identification, project proposal screening and policy development with regard to participatory IPM.

The Facility maintains, and draws upon, a global IPM network of national IPM programmes, NGOs, farmer associations, field trainers, research stations, international agricultural research centres, projects, international organisations, universities, foundations, donors and aid agencies. The network constantly presents new opportunities for information exchange, collaboration, projects and funding. Many of the opportunities can be realised with timely action by the Facility to identify and connect partners who can support each other.

Services provided by the Global IPM Facility are normally expected to be covered from project budgets. However, the following specific and free services are offered to its main partners, including NEDA:

- assistance in identification of new IPM project opportunities;
- review of proposals for projects in the field of plant protection;
- provision of inputs for policy documents.

Embassies can request such assistance through the IPM focal point at DGIS/DRU/RR in the Hague, or the Global IPM Facility by email at global-ipm@fao.org.

International collaboration

- **Encourage collaboration among developing countries**

Exchange visits and coordinating meetings among developing countries with national IPM programmes have been instrumental to the success of national IPM programmes based on participatory IPM. Embassy staff should continue to encourage such visits and meetings.

- **Encourage donor coordination at national level**

A growing number of donors and aid agencies are showing interest in participatory IPM. This means that there is a need for coordination. Coordination should be at two levels:

Firstly, at government level. Governments should be encouraged to establish a national IPM programme with an IPM Steering Committee of senior officials (e.g. head of plant protection; head of extension services; etc.). The national IPM programme should provide a coordinated framework for project activities. It should formulate an IPM policy and establish priorities and a time-frame for the introduction of IPM. Many Asian countries established such national IPM programmes and IPM Steering Committees. These committees now play an important coordinating role.

Secondly, donors should coordinate with each other, and with the Global IPM Facility, on how they can jointly and effectively support national IPM programmes.

Donor coordination is also important if donors are urged to provide pesticides for emergency operations to control migratory pests. Uncoordinated responses to such requests have led to excessive donations of products that eventually became obsolete and turned into hazardous waste causing severe environmental problems.

endnotes and references

1   Data adapted from: Handbook for incorporation of IPM in agricultural projects, Asian Development
     Bank, 1994; P.E. Kenmore, A perspective on IPM, ILEIA Newsletter, December 1997, Volume 13, No.
     4; Various reports of FAO-supported National IPM Programmes; data provided by the Global IPM
     Facility.

2   Panups 17/03/1997 referring to Agrow: World Crop Protection News, 13/12/1996, 14/02/1997 and
     28/02/1997.

3   Meerjarenplan Gewasbescherming; Lower House of the States-General; Vergaderjaar 1990-1991
     session.

4   Terminal Report: Intercountry Programme for the Development and Application of Integrated Pest Con-
     trol in Rice in South and South-East Asia, Phase I and II; Project Findings and Recommendations;
     Rome, 1994.

5   In Indonesia, with over 1 million FFS educated farmers in 1998, the total annual saving on farmer
     expenditure on pesticides is over USD 15 million, assuming an average plot size of 0.75 ha and 2 sea-
     sons per year. In Vietnam, where 400,000 farmers have passed through FFS, the annual saving exceeds
     USD 6 million. After having gained confidence in IPM as the main strategy for pest control in rice, the
     government of Indonesia abolished subsidies for rice pesticides, which saved USD 120 million per
     annum. The central government in India replaced subsidies on insecticides by a tax and is now gaining
     USD 60 million annually in savings and revenues, while annual spending on IPM field training is
     about USD 10 million.

6   M. Whitten and W.H. Settle (1998); The role of the small-scale farmer in preserving the link between
     biodiversity and sustainable agriculture.

7   Integrated Pest Management: farmer field schools generate sustainable practices, A case study in Central
     Java evaluating IPM training, E. van der Fliert, Wageningen Agric. Univ. Papers 93-3, 1993.

8   Mid-term review of Phase III of the FAO Intercountry Programme for the development and application
     of integrated pest control in rice in South and Southeast Asia, August 1996.

9   P.E. Kenmore; A perspective on IPM; ILEIA Newsletter; December 1997, Volume 13, No. 4.

10   Kremer, A.R. (1992); Pests and donors in Mali; the journal of disaster studies and management; Vol-
     ume 16, number 3, Blackwell publishers, Oxford, UK.

11   Joffe, S. (1995); Desert locust management: a time for change; Discussion Paper No 284: the World
     Bank; Washington.

12   Draft FAO paper with provisional title: economic and policy issues in desert locust economics: a prelimi-
     nary analysis. The paper is expected to be published in 1998.

[13] Our Stolen Future: How man-made chemicals are threatening our fertility, intelligence and survival; T. Colborn, J.P. Myers and D. Dumanoski; Little, Brown and Company; London; 1996, and various Panups during 1996 and 1997.

[14] FAO/UNEP/WHO Provisional technical guidelines on the disposal of bulk quantities of obsolete pesticides in developing countries, FAO, Rome, 1996.

[15] Public health impact of pesticides used in agriculture, WHO, Geneva, 1990; and Guidelines for aid agencies on pest and pesticide management, OECD/DAC Guidelines on Aid and Environment No. 6, OECD, Paris, 1995.

[16] The WHO-recommended classification of pesticides by hazard is commonly used for classification of pesticides by hazard. Hazard classes are based on LD 50 values. Products in Class Ia are regarded as extremely hazardous, Class Ib highly hazardous, Class II moderately hazardous and Class III slightly hazardous.

[17] Guidelines for aid agencies on pest and pesticide management, OECD/DAC Guidelines on Aid and Environment No. 6, OECD, Paris, 1995.

[18] An example: In 1995, 48,377 cases of poisoning were reported in China, 3204 of them fatal. Many of these cases were related to normal use pesticides listed in WHO Hazard Class I (Panups 14/10/97).

[19] Review of the implementation of the International Code of Conduct on the Distribution and Use of Pesticides, Report to the Thirteenth Session of the Committee on Agriculture, FAO, 1994. 59% (vs. 48% in 1986) of the responding developing countries state that labels do not conform with official recommendations; 63% (vs. 37% in 1986) state that unsubstantiated claims in advertising are sometimes or frequently a problem; 62% (vs. 57% in 1986) felt that the quality of pesticides as marketed was not the same as that cleared for registration.

[20] Mid-term review of Phase III of the FAO Intercountry Programme for the development and application of integrated pest control in rice in South and Southeast Asia, August 1996.

[21] S. Ange, G. Fleisher, F. Jungbluth, H. Waibel; Guidelines for Pesticide Policy Studies: A framework for analyzing economic and political factors of pesticide use in developing countries; a publication of the Pesticide Policy project (Publication Series No. 1), University of Hannover, Germany, 1995.

[22] E.g.: Bayer Agrochem Courier; Special Issue on Integrated Crop Management (1997). Formal statements on the necessity of IPM and the company's commitment to IPM are followed by a chapter entitled "overcoming prejudices about IPM" in which it is explained that IPM should be about optimising the use of agro-chemicals instead of minimising their use.

[23] A USD 400 bn gamble with world's food; Guardian Weekly, Vol, 125, No. 25 (1997).

[24] E.g.: GCPF website; news message about obtaining special NGO status at FAO (1997).

appendix 1

Abbreviations and acronyms

DAC	Development Assistance Committee of the OECD
DGIS	Directorate-General for International Cooperation of the Netherlands Ministry of Foreign Affairs
FAO	Food and Agriculture Organisation of the United Nations
FFS	Farmer Field School
GCPF	Global Crop Protection Federation (former GIFAP)
IPM	Integrated Pest Management
NGO	Non-Governmental Organisation
OECD	Organisation for Economic Cooperation and Development
TOT	Training of Trainers
T&V	Training and Visit (Specific approach to extension)
UNCED	United Nations Conference on Environment and Development
UNDP	United Nations Development Programme
UNEP	United Nations Environment Programme
WHO	World Health Organisation

appendix 2

Selected documentation

The following is a selection of publications on IPM and pesticide management in developing countries. The list is not meant to be exhaustive and the inclusion of a document in this list does not indicate that it is being recommended in preference to a document that is not listed.

Integrated Pest Management and sustainable agriculture

FAO (1998) *Community IPM: Six cases from Indonesia*; FAO Technical assistance Indonesian National IPM Program

ILEIA (1997) *Fighting back with IPM*; ILEIA Newsletter, December 1997

World Bank (1997) Integrated Pest Management: Strategies and policies for effective implementation; Environmentally sustainable development studies and monograph series, no. 13; the World Bank; Washington

UNDP (1996) *UNDP and Integrated Pest Management*: A Guiding Report by the Natural Resources Institute on behalf of UNDP.

Kenmore, P.E. (1996) *Integrated Pest Management in rice*, chapter in "Integrated Pest Management and biotechnology" by G.J. Persley, CABI.

Kenmore, P.E. (1995) *Indonesia's IPM - A Model for Asia*, FAO Inter-country Programme for Integrated Pest Control in Rice in South and South-East Asia, Manila.

Kenmore, P.E. *Empowering farmers: Experiences with Integrated Pest Management*;
Gallagher, K.D. and Entwicklung + Ländlicher Raum 1/95
Ooi, P.A.C. (1995)

OECD (1995) *Guidelines for Aid Agencies on Pest and Pesticide Management*, OECD Development Assistance Committee; Guidelines on Aid and Environment No. 6, Paris

Pretty, J.N. (1995) *Regenerating agriculture: Policies and practice for sustainability and self-reliance*; Earthscan, London

Waibel, H et al (1995) *Guidelines for Pesticide Policy Studies: A framework for analyzing economic and political factors of pesticide use in developing countries*; S. Agne, G. Fleisher, F. Jungbluth, H. Waibel; a publication of the Pesticide Policy project (Publication Series No. 1), University of Hannover, Germany

DGIS (1995)   *Biological diversity; Sectoral policy document of Development Cooperation 8; Ministry of Foreign Affairs, the Netherlands.*

ADB (1994)   *Handbook for incorporation of Integrated Pest Management in agricultural projects; Asian Development Bank, Manila*

DGIS (1993)   *Sustainable land use; Sectoral policy document 2; Development Cooperation, Ministry of Foreign Affairs, the Netherlands.*

Fliert, E. van der (1993)   *Integrated Pest Management: farmer field schools generate sustainable practices, A case study in Central Java evaluating IPM training,* Wageningen Agric. Univ. Papers 93-3.

UNCED (1992)   Agenda 21, Chapter 14 (Section I), *Integrated Pest Management and Control in Agriculture;* and Chapter 19, *Environmentally sound management of toxic chemicals.* Adopted on 14 June 1992, Rio de Janeiro

Internet sites   http://www.IPMnet.org
                            http://www.FAO.org

Pesticide Management:

FAO (1995)   *Provisional Guidelines: Prevention of Accumulation of Obsolete Pesticide Stocks,* FAO, Rome

USAID (1994)   *Bilateral Donor Agencies and the Environment: Pest and Pesticide Management;* prepared for USAID by the Environmental and Natural Resources Policy and Training Projects (EPAT) of the Winrock International Environmental Alliance, Arlington, Viginia, USA

World Bank (1994)   *Pesticide Policies in Developing Countries: Do they encourage excessive use?* World Bank Discussion Paper 238, prepared by Jumanah Farah

FAO (1991)   *The Initial Introduction and Subsequent Development of a Simple National Registration and Control Scheme,* FAO, Rome

FAO (1990)   *International Code of Conduct on the Distribution and Use of Pesticides* (amended version), Rome

FAO (1990)   *Personal Protection When Working with Pesticides in Tropical Climates,* FAO, Rome

FAO (1989)   *Environmental Criteria for the Registration of Pesticides* (Revised version); FAO, Rome.

WHO/IPCS   *Health and Safety Guides* (a continuing series of documents on the health and safety aspects of specific pesticides and other chemicals), International Programme on Chemical Safety, WHO, Geneva.

Useful web-sites :

WHO International Program on Chemical Safety (IPCS): http://www.who.ch (/programmes/pcs/pub_list.htm) or gopher://gopher.who.ch:70/11/.pcs/.ehc (Environmental and health information about specific pesticides)

Extoxnet: http://ace.ace.orst.edu/info/extoxnet/ghindex.html (detailed data sheets on specific pesticides)

Pesticide Action Network (PAN): http://www.panna.org/panna/ or gopher://gopher.igc.apc.org/11/orgs/panna/pestis (publications on pest and pesticide management: also possible to subscribe to PanUps, the newsletter of PAN)

Global Plant and Pest Information System: http://pppis.fao.org

appendix 3

Supporting participatory IPM: suggestions for Netherlands Development Cooperation staff

Type of project	Action	Points for special attention	References for further information
IPM projects	• support projects within the framework of the 4 stage model • support studies and small projects that contribute to a conducive policy environment (particularly pesticide policy analyses) • enhance donor coordination • keep abreast of developments in the country and take advantage of field days to visit projects	• the quality of the proposed IPM project in terms of technical content and the participatory process • possible bias towards chemical control or research • can use be made of experience gained in other countries?	• DGIS/DRU • Global IPM Facility • Wageningen Crop Protection Centre
Agricultural intensification projects including irrigation projects, rural or area development projects, commodity development projects, etc.	• review ongoing and future projects to identify possibilities for inclusion of participatory IPM components	• sustainability of production system • cost-effectiveness of production system • bias towards chemical control	• DGIS/DRU • Global IPM Facility • Wageningen Crop Protection Centre
Locust control activities	• evaluate necessity of chemical control • coordinate with other donors and FAO	• economic justification • environmental impact • risk of flooding local market with cheap pesticides	• DGIS • FAO Locust Group • Wageningen Crop Protection Centre
Requests for pesticide donations	• in principle, do not provide pesticides • use OECD checklist if pesticides are nevertheless supplied for any exceptional emergency • tie such emergency supplies to a formulation mission to determine more sustainable IPM-based solutions for the longer term	• bias towards chemical control by the institution requesting pesticides • likelihood of a negative impact on IPM programmes if pesticides are donated (form of subsidising pesticides and encouraging their use)	• DGIS/DRU • Global IPM Facility • OECD Checklist (see Appendix 4)
Other projects with an extension component that might benefit from the participatory approach developed under IPM (e.g. community forestry, natural resources management, women)	• investigate the scope for using the FFS approach in other sectors and enable relevant actors from projects in these sectors to familiarise themselves with FFS training	• learning through experimenting • ownership of the process	• DGIS/DRU • WAU: Group Communication and Innovation Studies

appendix 4

Ensuring good practices when providing pesticides under aid programmes (from the OECD Guidelines for aid agencies on pest and pesticide management)

Requests for the provision of pesticides should be subject to careful appraisal procedures to avoid inappropriate and excessive supplies. The following provides guidance for the appraisal of requests and for good practices when providing pesticides:

1. To enable a careful appraisal of requests for pesticide procurement, each request should be adequately justified and should be specific about: the intended use of the pesticides; the product specifications; the required quantity; packaging require- ments; items to be delivered with the pesticides in order to reduce hazards. The checklist (Box at the end of this Appendix) should be used to determine whether all necessary information has been provided. Missing information should be obtained through dialogue with the recipient country.

2. Pesticide procurement should fully comply with national pesticide legislation and regulations of the recipient country. The Prior Informed Consent procedure should be complied with.

3. Extremely and highly hazardous pesticides of WHO Class Ia and Ib and compounds which are highly persistent in the environment should not be provided. Exceptions can only be considered if all three of the following criteria are met: a) there are urgent reasons to use these pesticides, b) there are no safer alternatives, c) their safe and controlled application can be guaranteed. Pesticides of Class Ia, Ib and the more toxic range of Class II, are generally considered to be unsuitable for use by small-scale farmers.

4. The quantity provided should be consistent with the actual requirements and the local capacity to store, distribute and apply the pesticides. If any of these capacities is limited, the pesticides should be provided in smaller consignments spread over a period of time. Because of the limited shelf-life of most pesticides, stocks should not exceed their shelf life. The additional costs of providing pesticides in smaller consignments and different shipments will, in most cases, outweigh the costs that will be incurred if stocks become obsolete as a result of prolonged or improper storage.

5. Pesticides should only be provided in combination with "hazard-reduction pack- ages". Such packages should comprise the following items, unless it has been con- firmed that these items are already available in the area where the pesticides are intended to be used:
   - adequate quantities of antidotes with medical instruction for distribution to health posts in the area where pesticides that could cause serious poisoning are intended to be used,

- adequate quantities of appropriate protective gear,
- appropriate and safe application equipment,
- a number of salvage drums (over-drums to contain drums that start leaking) and a drum crusher with rinsing instructions for large quantities of pesticides packed in 200 litre drums (e.g. for locust control),
- Material Safety Data Sheets providing information on how to handle accidents with the pesticides concerned (available from the manufacturer of the product).

Further, the package should include training of users, both men and women, in the appropriate use of pesticides and in understanding hazards connected to the use of pesticides, if these have not yet been trained.

6. Containers should be durable enough to meet rough transport and storage conditions. Labels should be in the national language, as well as in the local language of the area of intended use, and have pictograms for illiterate users. Labels and packaging should comply with relevant FAO Guidelines. Purchase orders and tender documents should spell out the labelling requirements and the minimum container quality requirements.

7. The establishment of large on-site strategic stocks for locust control emergency operations should be avoided. Instead, donors should investigate the possibility of establishing so-called "pesticide bank" arrangements, whereby pesticides are kept stand-by at the location of their manufacture to be flown in when their use is actually required. Research into locust control methods which require less pesticides, should be given high priority.

8. Provision of large quantities of pesticides should be preceded by an environmental assessment. The environmental impact of extensive spraying operations under locust control emergency operations should be monitored.

9. Aid agencies are encouraged to assist recipient countries with the urgent problem of pesticide left-overs and obsolete pesticides through investigating possibilities for:
   - local refund-systems for empty containers of pesticides used by farmers;
   - arrangements with suppliers to accept unused left-overs of pesticides for reformulation or destruction.

10. Aid agencies are encouraged to establish internal criteria and procedures for the selection and procurement of pesticides. The FAO provisional Guidelines on the Tender Procedures for the Procurement of Pesticides provide guidance in this field.

Checklist: **information required for appraisal of requests for the provision of pesticides**

Use
- purpose for which the pesticides are required,
- reason why the use of pesticides is necessary and alternative non-chemical methods cannot be used,

Product specifications
- justification for the selection of the requested active ingredient and formulation (referring to: efficacy; environmental considerations; occupational and public health considerations; the available type of application equipment),

Quantity
- required quantity (referring to the extent of infestation and the size of the area to be treated; present stocks; capacity to distribute the pesticides effectively; application capacity in terms of available equipment and trained staff; storage capacity),

Packaging requirements
- required quality of packaging (referring to: climate; storage and transport conditions; risk of prolonged storage),
- required package size (referring to the available type of application equipment),
- required languages for labels,

Hazard reduction
- level of knowledge among envisaged users about the hazards connected to the use of pesticides and their appropriate use (to determine whether training is necessary),
- availability of protective gear in area of use (to determine whether protective gear should be supplied with the pesticides),
- availability of antidotes at health posts in the area of use (to determine whether antidotes should be supplied with the pesticides), availability of facilities to dispose of empty containers (to determine whether a drum crusher should be supplied with the pesticides).